Critical Facilities Engineering:

A Comprehensive Training Guide

Yekini K Tidjani

ISBN

Hardcover: 979 8 9989129 3 1

Paperback: 979 8-9989129 2 4

Dedication

This book is dedicated to the unsung heroes the data center industry the vital facilities engineers who work tirelessly behind the scenes to keep the digital infrastructure running seamlessly. Your dedication, expertise, and commitment to maintaining uptime are genuinely inspiring.

This work honors your dedication and highlights the essential contribution you make to the success of many businesses and organizations. It is also dedicated to aspiring individuals seeking to enter this important field, offering a pathway to success in a dynamic and fulfilling industry.

May this guide provide you with the skills knowledge to succeed, make meaningful contributions, and develop a successful career in supporting our modern digital infrastructure. Your dedication to maintaining reliable and efficient data center operations fuels the innovation and connectivity that sustain our daily lives. This is dedicated to you – the past, present, and future generations of critical facilities engineers.

Table of Contents

Chapter 1: Introduction to Critical Facilities Engineering in Data Centers

Understanding Modern Data Centers

Managing a modern data center involves more than just handling rows of equipment; it's about orchestrating a dynamic, interconnected system. Every component, from cooling and power to networking and layout, impacts the facility's stability and efficiency. A critical facilities engineer serves as the conductor of this complex operation, ensuring all parts work harmoniously and stepping in to address any issues that arise.

At the heart of the operation is the data hall. This is where the action happens—rows of server racks packed with computing power, processing petabytes of data around the clock. The positioning these racks is no accident. The layout follows the hot aisle/cold aisle containment principle: cold air is channeled into the front of the servers, hot exhaust air is directed away from the back to prevent overheating. This simple concept has a massive impact on energy efficiency and equipment longevity.

Each detail from the density of the servers to the type of cables used affects energy use and cooling demands. And that brings us to the next vital component: the network.

The network infrastructure is the circulatory system of the data center. Fiber optic cables, floors, and trays, linking switches, routers, and firewalls to create an ultra-fast, high-capacity environment where data flows seamlessly. Redundancy is built into every design, ensuring no single point of failure bring the system down.

Here's a quick breakdown of common network topologies and how they compare:

Network Topologies Comparison

Topology	Advantages	Disadvantages	Typical Use Case
Star	Easy to install and manage; failure of one device doesn't affect others	Central hub failure disables the network	Office LANs, small data centers
Mesh	Highly reliable; multiple paths	Expensive and complex to	Military systems, high-
	ensure no single point of failure	install and maintain	reliability applications
Ring	Data travels in one direction,	Failure in one device can affect the entire	Legacy systems, simple token

	easy to manage and isolate faults	network unless dual ring	ring networks

Of course, all this technology requires a significant amount of power. That's where the power distribution system comes in. It begins with the utility grid and flows through several layers of protection and transformation, including transformers, switchgear, UPS (Uninterruptible Power Supply) systems, and ultimately, Power Distribution Units (PDUs), before reaching the server racks.

Here's how that journey looks:

Power Flow in a Data Center

Each element plays a role in converting and safeguarding power delivery. The UPS bridges the gap during short-term outages, and diesel generators cover extended blackouts. Engineers monitor power usage in real time, balance loads across circuits, and anticipate spikes that could threaten uptime.

Additionally, working with high voltage systems requires strict adherence to safety protocols, especially lockout/tagout (LOTO) procedures, which prevent accidental re-energization during maintenance and other operations.

Capacity planning is just as important. Overloading the

power infrastructure can lead to significant failures. Engineers must analyze real-time data from power meters and software to ensure the entire system operates within safe and efficient limits.

Cooling Systems, Security, and System Interdependencies

Cooling systems are one of the most mission-critical components in a data center. With densely packed IT equipment generating significant heat, proper cooling isn't optional—it's essential. Without it, servers can overheat, leading to system failures, data loss, and serious downtime costs.

To manage this, data centers use a variety of cooling technologies. The core strategies typically include Computer Room Air Conditioners (CRACs) or Computer Room Air Handlers (CRAHs), along with Fan Wall Units (FWUs) or Fan Coil Units (FCUs). These systems circulate cool air throughout the facility, ensuring all server equipment remains within safe temperature thresholds.

In larger operations, chiller systems provide chilled water to CRAC units for more effective cooling. Engineers should also understand advanced systems, such as adiabatic cooling, free cooling, and liquid cooling, each offering distinct advantages and operational factors.

Cooling Technologies Comparison

Technology	Best For	Key Advantages	Considerations
CRAC/CRAH	Standard data center cooling	Reliable and widely used	Regular maintenance needed
Chiller Systems	Large-scale, centralized facilities	Scalable and efficient	Higher upfront cost and complexity
Adiabatic Cooling	Dry climates with water availability	Energy-efficient and eco-friendly	Requires water and climate suitability
Free Cooling	Cool ambient climates	Low operational costs	Dependent on the external climate

Liquid Cooling	High-density, high-performance computing	Superior thermal efficiency	Higher infrastructure investment

Security, too, is fundamental. Data centers hold not only valuable hardware but sensitive digital assets. That's why both physical and logical security systems must work in unison. Physical systems include CCTV, biometric access control, and intrusion detection. Logical protect networks with firewalls, access rules, and cybersecurity protocols.

The facilities engineer plays a vital role in monitoring, responding to, maintaining these on a daily basis. They must also understand how to interpret security data, enforce access controls, and ensure compliance with standards like HIPAA and PCI DSS.

These systems don't exist in isolation. They are deeply interconnected. For example, a power outage may lead to a cooling failure, triggering a cascade overheating, server failure, and possibly a false alarm in the security system. Conversely, a security breach might require coordinated emergency power and cooling shutdowns.

That's why adopting a holistic approach is crucial. Engineers must develop a comprehensive understanding of these interdependencies, establish effective communication channels with IT teams, tenants, and vendors, and remain proactive with a focus on prevention and preparedness.

Proactive maintenance, real-time monitoring, and scheduled system reviews are not just tasks they're pillars of successful operations. From managing vendor relationships to inspecting equipment to analyzing data trends, the facilities engineer wears many hats, all of which directly affect uptime and system resilience.

In tenant-occupied data centers, this role becomes even more dynamic. Here, the engineer also serves as a key liaison, balancing needs of occupants with the physical infrastructure's constraints. It's a role that blends technical acumen with diplomacy, making communication and collaboration just as crucial as troubleshooting and planning.

Preventative Maintenance and Operational Duties

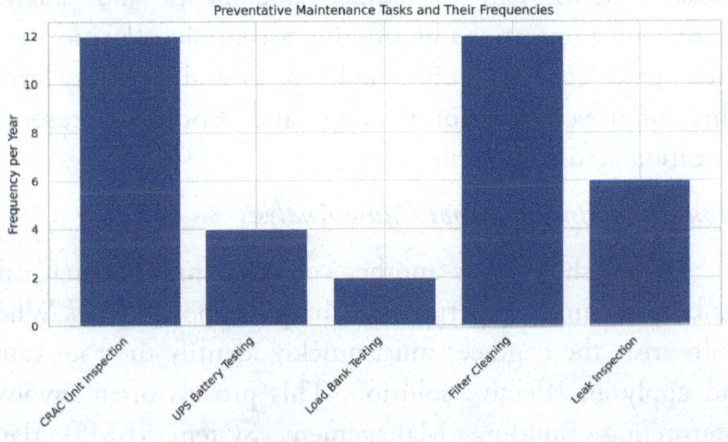

Preventive maintenance is fundamental to a critical facilities engineer's role. It is a proactive approach aimed at preventing failures rather than reacting to them. Achieving success in this area depends on accurate scheduling, careful adherence to manufacturer specifications, and a deep understanding of each system's operation.

For instance, regular CRAC unit inspections include checking refrigerant levels, testing fan functionality, inspecting filters, and identifying and sealing leaks. Similarly, UPS systems demand battery testing, load bank exercises to verify capacity, and checks on the charging infrastructure. These procedures significantly reduce the risk of downtime, supporting the uninterrupted operation of essential systems.

Every maintenance activity is carefully logged, noting the date, time, actions performed, and key findings. This recordkeeping is vital for spotting performance trends, understanding recurring issues, and predicting future needs. To manage all of this efficiently, facilities engineers rely on Computerized Maintenance Management Systems (CMMS). These systems help schedule, document, and analyze maintenance operations by offering a centralized platform for tracking work orders, equipment histories, and inventory levels. This improves decision making and optimizes resource allocation across the facility.

Troubleshooting Root Cause Analysis

Troubleshooting is another core responsibility that calls for both technical expertise and sharp diagnostic skills. When faults arise, the engineer must quickly identify the root cause and apply an effective solution. This process often involves interpreting Building Management System (BMS) data, reviewing system logs, and performing on site inspections.

For example, if there is a sudden loss in cooling, potential causes might include a failing chiller, a CRAC malfunction, or an airflow obstruction. The engineer must systematically investigate each possibility using tools like infrared cameras, diagnostic software, and airflow meters. The troubleshooting process is frequently collaborative, requiring assistance from external vendors or specialized technicians. Each incident is followed by a written Root Cause Analysis (RCA) report, which documents the issue, actions taken, and resolutions. The RCA reports contribute to continuous improvement by helping teams avoid repeated failures and reinforcing knowledge through experience.

Troubleshooting and Root Cause Analysis Workflow

Emergency Preparedness Response

Emergency response is one of the most critical functions of a facilities engineer. During high stress situations such as power loss, fire alarms, or HVAC system failures the engineer is the first responder. This role requires deep knowledge of emergency procedures, calm execution under pressure, and effective coordination with other teams. Emergency protocols must be predefined, regularly rehearsed, and readily available to all personnel.

Consider a power outage scenario: engineers must immediately transfer loads to backup systems, confirm UPS operation, and inform affected tenants with updates restoration estimates. In the event a fire alarm, steps may include initiating evacuations, coordinating with emergency responders, and assessing the impact on operations. Post-incident reviews vital for capturing lessons learned and improving future responses.

Vendor Oversight and Relationship Management

Vendor management is a significant aspect of the role. It involves supervising contractors, managing service agreements, and ensuring that vendors meet performance and safety standards. Engineers must maintain proactive communication, schedule service appointments, and ensure all work is completed according to scope. Vendor selection is a strategic decision based on technical competence, reliability, cost, and responsiveness. The facilities engineer is responsible for setting expectations, monitoring performance, and ensuring compliance with contract terms. Regular check ins and performance evaluations help maintain accountability and provide ongoing progress. Additionally, strict access controls, background checks, and compliance with security protocols ensure that vendors do not introduce risk to the data center.

Tenant Support and Service Coordination

Tenant relations are central to operations in a multi-tenant data center. The critical facilities engineer serves as the liaison between facility systems and the individuals who rely on them. This means understanding tenant needs, providing clear communication, and ensuring timely responses to service-impacting events. For instance, a tenant with network performance issues may require the engineer to coordinate between facility networking, tenant IT teams, and third party service providers.

Engineers often need to translate technical details into clear language, update tenants on progress, and manage expectations with professionalism and transparency. Establishing trust and maintaining a strong working relationship helps ensure satisfaction and smooth day to day operations.

Reporting Cross-Functional Communication

Effective communication is essential across all aspects of the engineer's role. Engineers routinely prepare performance summaries, maintenance logs, and incident reports for stakeholders, including management, tenants, and external partners. Whether through written documentation, emails, meetings, or dashboards, clarity precision in communication are key to maintaining operational awareness and stakeholder confidence.

The communication method depends on the situation—urgent issues may require phone calls instant messages, while routine updates may be delivered through weekly reports or scheduled meetings. Strong communication also promotes transparency and proactive collaboration across teams.

Safety and Lockout/Tagout (LOTO) Protocols

A strong culture of safety is non-negotiable in data center environments. Even one mistake can result in equipment damage, downtime, or injury. The critical facilities engineer must strictly follow established safety standards and procedures at all times.

Electrical safety relies heavily on Lockout/Tagout (LOTO) procedures. Before working on energized systems, the equipment must be fully powered down. A lock is placed on the switch to prevent it from being re-energized, and a tag is attached with information about who applied lock, when it was used, and the reason for its application.

If multiple personnel involved, each apply their lock and tag. Importantly, only the person who placed the lock may remove it, ensuring full accountability and safety.

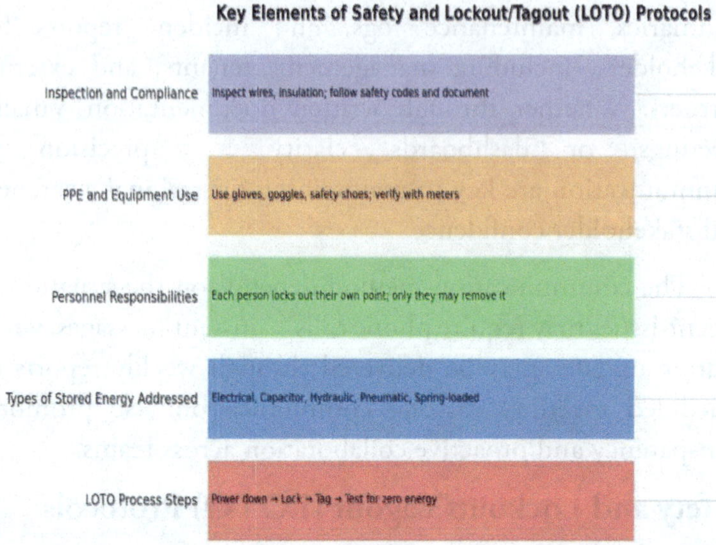

The LOTO process extends beyond simply locking out the power. It encompasses all forms of stored energy, including capacitors, springs, hydraulic systems, and pneumatic systems. Each type of stored energy requires specific procedures to ensure complete de energization. For example, a capacitor bank may require additional steps, such as discharging the capacitors using a specialized tool, before applying the lockout device.

Similar meticulous steps are necessary when dealing with pneumatic and hydraulic systems, ensuring that all pressure is released before commencing any work. Detailed procedures for each piece of equipment should be documented, readily accessible to all personnel, and updated whenever changes to the equipment or its maintenance procedures occur. Regular training sessions practical demonstrations crucial to ensure that everyone thoroughly understands and can correctly execute the LOTO procedure. Failure to follow these procedures meticulously can result in severe injuries and significant operational disruption.

Electrical safety extends beyond LOTO procedures. Data centers house a vast array of electrical equipment, operating at various voltage levels and currents. Personnel must be adequately trained and equipped to work safely around this equipment. This includes wearing appropriate Personal Protective Equipment (PPE), such as insulated gloves, safety glasses, and safety shoes. Additionally, regular inspections of electrical equipment are crucial to identify potential hazards, such as frayed wires, damaged insulation, and loose connections.

These inspections should be performed by qualified personnel, following established procedures documenting their findings. The use electrical testing equipment, such as multimeters and insulation testers, is crucial to ensure the integrity of electrical systems and identify potential hazards before they lead to incidents. Furthermore, a clear understanding electrical safety regulations standards is crucial for all personnel working in the data center.

These regulations often dictate the requirements for electrical safety inspections, personnel training, and the use of specific safety equipment.

Confined Space Safety and Emergency Preparedness

Confined space entry procedures are crucial in data center environments, where staff may need to access enclosed areas such as cable trenches, electrical vaults, underfloor systems. These areas present unique risks, including reduced oxygen levels, toxic gas buildup, and potential entrapment. Before entering any confined space, a complete hazard assessment must be conducted. This includes testing for adequate oxygen, checking for harmful gases, and confirming that ventilation is sufficient. In many cases, specialized tools such as multi gas detectors and ventilation blowers are required to ensure safety.

A permit to work system is often used to formalize entry into confined spaces. This ensures that all hazards are clearly identified, mitigation steps are in place, and the entry is supervised by trained personnel.

The permit includes a summary of risks, safety measures to be taken, and detailed emergency instructions. A designated individual is responsible for monitoring conditions and maintaining constant contact with those working inside the space. Should an emergency arise, this individual is responsible for initiating a rapid response and immediately alerting emergency services.

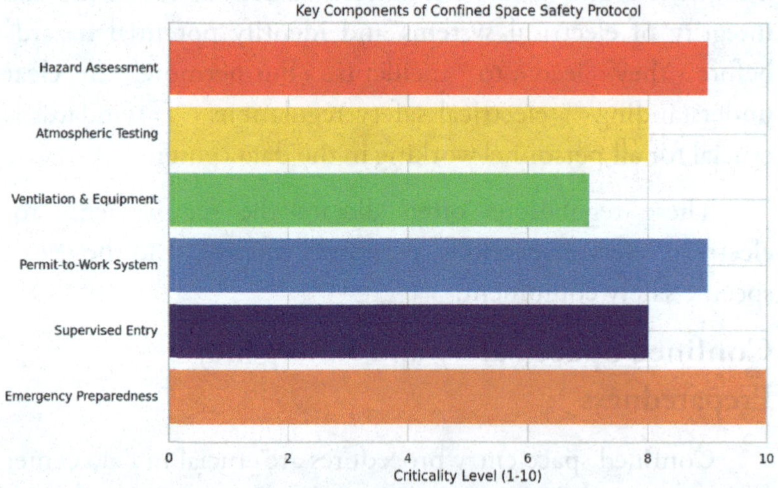

Emergency Response Coordination

Robust emergency response planning is crucial for protecting personnel and critical infrastructure in the event of unplanned incidents. These plans should outline step by step procedures for various scenarios, including fire, power failure, or medical emergencies. Conducting regular drills and training ensures that all staff are familiar with the procedures and can act quickly when needed.

Emergency protocols must be clearly written, easily accessible, and regularly reviewed to ensure their effectiveness and continued relevance. Importantly, a data center's emergency response plan should not function in isolation. It must be integrated with the larger emergency framework of the building, campus, or even the surrounding municipality. This integration facilitates coordinated response actions clear lines of communication across all involved parties. The plan should define specific roles responsibilities to reduce confusion during an actual event. Following any emergency, a formal review is necessary to assess the effectiveness of the plan identify areas for improvement to enhance future readiness.

Compliance Regulatory Requirements

Safety in a data center also entails adhering to all relevant legal and regulatory standards. These may include laws relating to occupational health and safety, fire prevention, and environmental protection. The exact requirements can vary based on jurisdiction, but compliance is non negotiable. Engineers critical facility managers stay informed about updates to regulations and ensure that any procedural or documentation changes are implemented swiftly.

The consequences of non compliance can be severe, ranging from financial penalties and legal action to reputational damage. To maintain compliance, regular audits safety inspections are necessary. These reviews verify whether operational procedures meet established standards, assess the reliability of emergency protocols, and confirm that staff are properly trained and equipped to follow safety rules.

Building a Culture of Safety

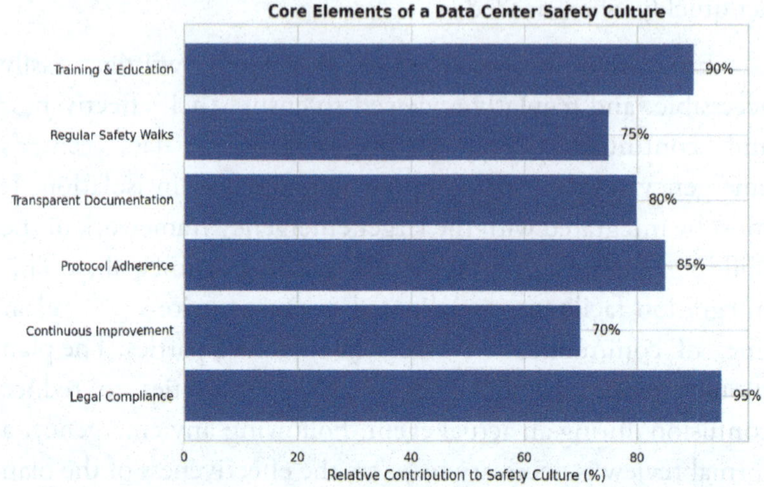

Safety is more than a checklist—it's a mindset. The success of a data center depends on how seriously everyone prioritizes safety at all levels. Following established protocols, complying with legal requirements, promoting a proactive safety culture all contribute to creating a secure and stable workspace. This involves providing thorough training, performing regular safety walks and inspections, and keeping transparent records of all activities.

When safety becomes an integral part daily operations, it leads to a more productive and resilient facility. The result is fewer incidents, improved staff confidence, and stronger protection for infrastructure assets. A culture continuous improvement where feedback is used to enhance procedures helps ensure that safety standards evolve alongside the facility itself.

Electrical Systems: The Power Backbone of the Data Center

At the core of every data center is its electrical infrastructure. Understanding this system is crucial for anyone involved in the operation of critical facilities. This section outlines power path from the external utility grid to the individual server racks, focusing on real world applications rather than complex technical theories.

The process begins with power delivery from the external utility grid. This incoming electricity typically arrives at medium voltage (such as 34.5 kV) is directed to an on site substation. There, large power transformers reduce the voltage to more usable levels, commonly around 480 volts. These transformers are designed for continuous operation and often use oil cooling to manage heat. Periodic maintenance, including oil sampling and electrical testing, is required to ensure reliable function. A transformer failure at this stage could result in a complete loss power across the facility.

From the substation, power is distributed through an organized and redundant network of switchgear, insulated cables, and busways. Switchgear systems serve as the control centers for electrical routing, enabling safe and intelligent distribution to various components within a data center. Today's switchgear often includes smart monitoring capabilities that support remote operation automated fault detection. Redundancy is built into the design, allowing for a seamless transition if one part of the system fails, thereby minimizing downtime.

Power flows across copper conductors housed in fire-rated insulated cable systems. These cables run neatly along cable trays through conduits to ensure safety comply with local code requirements. Their layout is carefully designed to support both electrical efficiency and fire containment.

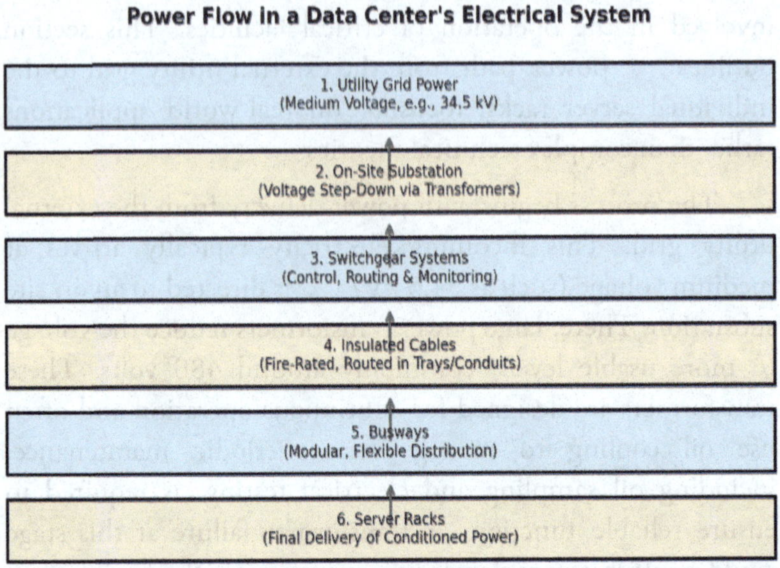

Busways complement cable systems by offering flexibility in distributing power along raised floors ceilings. These enclosed conductor systems modular easily reconfigured, which is especially useful in facilities with changing power needs or rapid growth. Busways make it easier to supply and reroute electricity to server racks without major construction or disruption to operations.

UPS Systems, Generators, and Power Optimization in Data Centers

A vital component in a data center's electrical infrastructure is the Uninterruptible Power Supply (UPS). UPS serve as the primary line of defense against power loss, providing immediate backup power to keep critical IT systems running uninterrupted. There are different types of UPS architectures, each tailored to specific operational needs.

Online (double conversion) UPS systems offer the highest level of protection by constantly converting incoming AC power to DC, and then back to AC, delivering a clean and consistent power output regardless of utility conditions. In contrast, Offline or Standby UPS systems only engage when utility power fails, making them more cost-effective but less robust for sensitive applications. The selection of a UPS system depends on the criticality of the protected load and the level of redundancy required. Proper sizing is crucial. A UPS must support the peak power demands of the IT infrastructure while also accounting for efficiency, future growth, and scheduled maintenance. Regular servicing including battery testing, replacement scheduling, and inverter checks is essential to ensure reliable performance during outages.

Generators: Long-Duration Backup

Generators add an extra level of power security, particularly during extended utility outages. Most data centers rely on diesel generators that can support the entire facility load for extended periods. These generators are typically connected to automatic transfer switches (ATS), ensuring a smooth power transfer in the event of an outage.

To ensure operational readiness, generators must be regularly tested under load, and maintenance schedules should include fuel system checks, oil changes, coolant tests, and exhaust inspections.

Fuel storage and resupply logistics require careful planning, especially for facilities in remote or regulated regions where environmental standards emergency delivery deadlines are critical. Generator capacity should incorporate a safety margin to accommodate unforeseen surges or increased power demands.

Real-Time Monitoring DCIM Integration

Monitoring holds equal importance to the electrical design. Data Center Infrastructure Management (DCIM) platforms, together with specialized monitoring tools, provide detailed, real-time insights into power consumption, circuit loads, voltage levels, and system performance. They generate alerts for anomalies such as current spikes or phase imbalances enabling engineers to resolve issues early and prevent failures.

Beyond alerting, monitoring tools enable predictive maintenance. For example, they can track battery degradation over time, identify trends in load usage, and highlight equipment at risk of failure. These insights help optimize energy efficiency, enhance planning accuracy, and support more informed capacity management. Engineers can use historical data to make informed decisions about upgrades, replacements, or procedural changes.

Power Factor Correction Efficiency

Modern IT equipment often introduces non linear loads, which can degrade the power factor increase strain on the electrical system. A low power factor results in inefficient energy use and higher utility costs. To counter this, facilities install power factor correction (PFC) capacitors, which help align the phase difference between current and voltage, reducing losses and stabilizing the power system. Implementing PFC not only leads to cost savings but also minimizes heat generation, reduces electrical noise, and improves the lifespan of power distribution components. It's a crucial step in optimizing overall infrastructure efficiency.

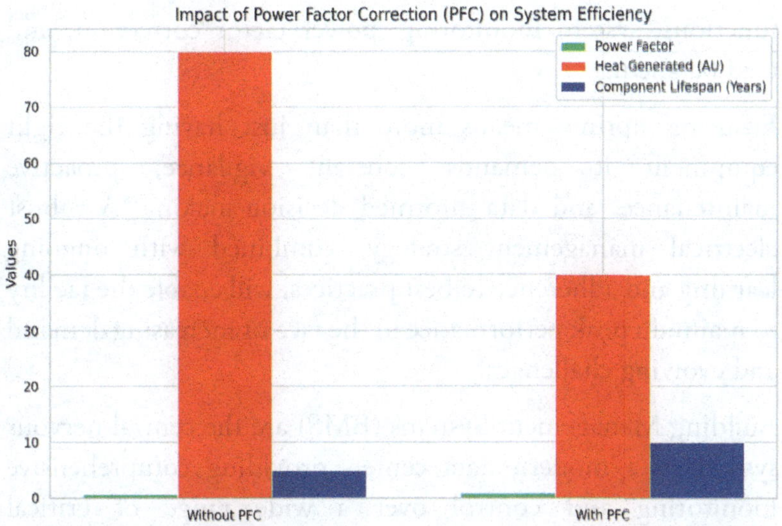

Load Balancing Across Power Distribution Systems

Effective load balancing ensures that electrical loads are evenly spread across Power Distribution Units (PDUs), circuit panels, UPS systems. An unbalanced load cause certain circuits to carry excess current, increasing the risk of overheating and failure. Over time, this imbalance also leads to reduced lifespan and unnecessary wear on power components.

DCIM tools often provide graphical interfaces to visualize load distribution and track trends over time. Engineers can utilize this data to reconfigure circuits, strategically shift loads, and plan for expansion without overstressing any part of the system. Proper load balancing is not a one time event—it requires ongoing adjustment as IT demands evolve.

Bringing It All Together

In summary, critical facilities engineers must master the intricacies of a data center's electrical system. This includes understanding the full power chain, from utility feed to server rack, as well as in-depth knowledge of UPS setups, generator functions, system monitoring, power factor correction, and load balancing.

Ensuring uptime means more than just having the right equipment it demands constant vigilance, proactive maintenance, and data informed decision making. A robust electrical management strategy, combined with ongoing learning and adherence to best practices, will enable the facility to maintain peak performance in the face of increasing demand and evolving challenges.

Building Management Systems (BMS) are the central nervous system of a modern data center, providing comprehensive monitoring and control over a wide range of critical infrastructure components. Understanding their functionality is paramount for any critical facilities engineer. These systems aggregate data from diverse sources, including HVAC systems, power distribution units (PDUs), security systems, fire suppression systems, and even lighting, providing a single, unified view of the data center's operational status. This integrated approach facilitates efficient management, proactive maintenance, and rapid response to potential problems.

A typical BMS architecture consists of several key components. At its core lies the central management station, a server or workstation running specialized software that collects, processes, and displays data from various field devices. These field devices, which include sensors and actuators, are scattered throughout the data center and are responsible for collecting data (such as temperature, humidity, and power consumption) and executing control actions (adjusting thermostat setpoints, opening and closing dampers, etc.). Communication between these field devices and the central management station occurs over a network, often utilizing various communication protocols such as BACnet, Modbus, or LonWorks. The choice of protocol often depends on the specific equipment being monitored and controlled, as well as the existing infrastructure in place. Understanding these protocols their interactions is crucial for troubleshooting and system integration.

The BMS interface, the point of interaction for human operators, typically consists a sophisticated software application providing a user friendly dashboard.

This interface often displays critical parameters in real time, using graphical representations such as trend charts, gauges, and maps, to facilitate easy comprehension complex data sets. The level of detail and complexity varies depending on the size and sophistication of the data center. A small data center may only require a basic interface displaying key parameters, such as temperature humidity. At the same time, a large facility may have a more elaborate system with multiple screens, drill-down capabilities, and advanced reporting tools.

The interface is not merely a display; it serves as a control center, enabling operators to adjust setpoints, initiate alarms, and schedule maintenance activities. Effective use of the BMS interface requires a high degree understanding, not only of the software itself but also of the underlying systems it controls. Training and experience are essential for navigating the complexities and interpreting data effectively.

Mastering BMS Data Interpretation in Critical Facilities Engineering

Interpreting data from a Building Management System (BMS) is a vital capability for any critical facilities engineer. Modern BMS platforms generate enormous volumes of real-time and historical data across numerous parameters, making the ability to extract meaningful insights essential. This goes far beyond simply monitoring alarms it involves detecting trends, anomalies, and subtle inefficiencies that could indicate larger system issues. For example, a gradual increase in Power Usage Effectiveness (PUE) over several weeks may suggest growing inefficiency in cooling or power delivery. Similarly, a sudden rise in chilled water temperature may indicate a potential failure in the cooling loop. Detecting these early signals enables preventative action, distinguishing experienced engineers from those who only respond to immediate issues.

The BMS also provides access to highly detailed and granular data that supports informed strategic decision-making. For instance, by reviewing individual Power Distribution Unit (PDU) readings, engineers can identify underutilized or overloaded server racks. This data informs load balancing decisions, helps optimize energy consumption, and can even guide the placement of equipment. Similarly, tracking temperature and humidity inside specific cabinets can identify thermal hotspots and enable targeted cooling adjustments, thereby improving performance and reducing unnecessary energy expenditure. These insights contribute directly to improving efficiency, redundancy, and operational integrity.

Proactive maintenance is greatly improved by analyzing historical data trends collected through the BMS. Instead of waiting to perform repairs after faults happen, engineers can plan maintenance based on predictive signals. For instance, monitoring vibration patterns of pumps or chillers can detect early signs of mechanical stress or bearing issues. Spotting these problems early allows for timely action, minimizing downtime and preventing expensive emergency repairs. This proactive strategy lowers operating costs, maximizes system uptime, and supports sustainable operation through efficient resource management.

Troubleshooting is another area where BMS data proves extremely helpful. When unexpected events happen, such as a sudden drop in server room temperature or an unexplained spike in power usage, the BMS offers a complete record of conditions across all systems. For instance, if room temperatures suddenly fall, engineers might initially suspect a cooling system failure. However, a closer look at BMS data could identify a sensor calibration mistake or a false reading, which helps refocus troubleshooting and prevents unnecessary repairs. Access to this detailed diagnostic information enables problems to be resolved quickly and accurately, reducing their impact on operations.

A practical example illustrates this well: suppose the BMS logs a slow but steady increase in the temperature within a specific hot aisle containment area. The engineer doesn't stop at just noting the trend they dig deeper. They begin by verifying airflow through aisle fans, checking for obstructions or failures. If the fans appear functional, attention shifts upstream to CRAC units, FCUs, or FWUs supplying chilled air. By cross referencing airflow rates, chilled water temperatures, and cooling unit runtimes, engineers can narrow down the exact point of failure. This layered analysis ensures corrective actions are accurate and efficient, minimizing downtime and wasted effort.

Efficient use of the BMS also requires familiarity with its interface and navigation. BMS platforms typically offer a tiered access structure, providing different levels of visibility and control based on user role. Engineers must be proficient in navigating menus, filtering data, generating custom reports, and interpreting trend graphs to manage data effectively. Hands on training such as drills that involve locating specific data points or simulating system faults is key to mastering these skills. In emergency scenarios, the ability to rapidly locate relevant data can make the difference between swift recovery and prolonged outages.

In summary, the Building Management System is far more than a passive monitoring tool it serves as the command center for the data center's infrastructure. A skilled engineer leverages the BMS not only to detect and respond to issues but also to optimize performance, improve planning, and support long term strategic goals. Mastery of the BMS involves understanding its structure, communication protocols, and data hierarchy, as well as applying critical thinking to interpret complex data patterns. Through training, experience, and continuous engagement with the system, critical facilities engineers elevate themselves from passive operators to proactive stewards of reliability and efficiency. This mastery directly supports improved uptime, smarter energy use, and long term cost savings core priorities in any successful data center.

Chapter 2: HVAC Systems in Data Centers

Understanding Cooling Systems in Data Centers

Efficient heat removal is essential for the continuous and dependable operation of a data center. IT equipment produces substantial heat, and without adequate cooling, system performance and uptime are quickly threatened. Critical facilities engineers need to fully understand the types of cooling systems available, how they function, and the common maintenance and troubleshooting issues they pose. This section highlights the most commonly used cooling technologies, emphasizing real world applications and best practices for operation.

CRAC Units: The Foundation of Air-Based Cooling

Computer Room Air Conditioners (CRACs) are foundational to cooling in many traditional data centers. These self contained units rely on a standard refrigeration cycle, typically using refrigerants such as R 407a or newer, eco-friendlier alternatives. Heat is absorbed from the room air and expelled to the external environment.

A standard CRAC unit consists of several core components:

- **Evaporator Coil** – where warm data center air transfers its heat to the refrigerant.

- **Compressor** – increases the refrigerant's pressure and temperature.
- **Condenser Coil** – releases heat to the external air.
- **Expansion Valve** – regulates the flow and pressure of refrigerant within the system.

Fans within the CRAC unit draw in warm air from the room, circulate it across the evaporator coil for heat exchange, and then return the cooled air into the environment. Proper airflow design is crucial to prevent inefficiencies and ensure consistent cooling throughout the space.

CRAC units must be properly sized to match the data center's thermal load. Oversized units can lead to energy waste due to short cycling, while undersized units risk overheating the IT environment. Design factors include room dimensions, IT rack density, and ambient conditions.

Strategic placement within hot and cold aisle configurations supports optimal air distribution and prevents hot spots. Regular maintenance, including filter replacement, coil cleaning, and refrigerant level checks, is essential to ensure maximum efficiency and prolong equipment life. Neglected maintenance can quickly degrade performance and lead to system failures.

CRAH Units: Chilled Water Efficiency

Computer Room Air Handlers (CRAHs) offer a scalable alternative to CRAC systems. Instead of relying on internal refrigerant loops, CRAHs are connected to a centralized chilled water plant, using this external supply to remove heat from the server room.

Warm air passes through the unit and across a coil filled with chilled water. The heat is transferred to the water, which is then returned to the chiller for cooling. This closed loop system supports centralized control and often improves energy efficiency in large scale operations.

CRAH design considerations are similar to CRACs but require careful coordination with the chilled water system to ensure proper sizing and flow rates. The effectiveness of a CRAH setup depends on maintaining consistent water flow, setting optimal temperature setpoints, and conducting continuous monitoring.

Maintenance tasks include changing filters, cleaning coils, and checking chilled water valves and connections for signs of leaks or flow issues.

Choosing Between CRAC and CRAH Systems

Selecting between CRAC and CRAH systems depends on data center size, infrastructure budget, and long-term scalability needs:

- Small to medium data centers often favor CRACs due to their standalone operation and lower upfront cost.
- Large facilities typically benefit from CRAHs integrated into a centralized chilled water plant for improved efficiency and modular growth.

Many facilities adopt a hybrid approach, combining both technologies to achieve optimal results. This allows for redundancy if one system fails, the other can maintain acceptable cooling conditions and supports dynamic load balancing during maintenance or temperature fluctuations.

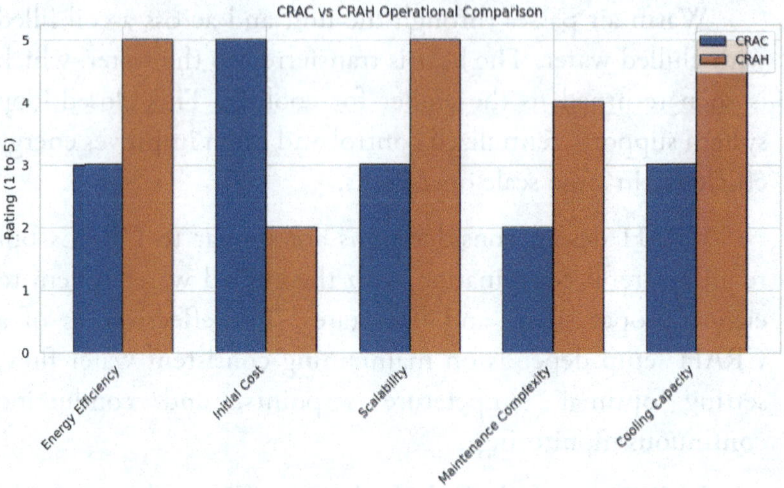

Liquid Cooling: Direct and Immersive Technologies

With increasing rack densities and thermal loads, liquid cooling systems are gaining popularity for their high efficiency and compact design. These systems provide direct contact cooling, either at the component or server level.

- **Direct-to-Chip Cooling:** Uses conductive metal blocks to apply coolant directly to CPUs and GPUs. This method removes heat at its source, enabling better thermal performance, lower fan speeds, and quieter operation.
- **Immersion Cooling:** Entire server components are submerged in a dielectric (non conductive) cooling fluid. Heat is absorbed directly by the liquid, which is then circulated through heat exchangers or chillers. This technique enables extremely high cooling

densities, often required in AI, HPC, or cryptocurrency mining environments.

Liquid cooling presents notable benefits improved efficiency, compact form factor, and quiet operation but also introduces unique challenges. Maintenance requires leak proof plumbing, robust leak detection systems, and specialized coolant fluids that are safe, stable, and compatible with IT hardware. Personnel must be trained in handling procedures, including safe servicing and disposal of coolant.

Cooling System Optimization and Troubleshooting in Data Centers

Selecting the right cooling technology is a strategic decision that directly influences operational efficiency and long term energy costs in a data center. Factors such as IT load density, physical infrastructure, environmental conditions, and budget constraints must all be carefully evaluated. An effective cooling strategy also requires a thorough understanding of airflow management techniques. Systems like hot aisle containment and cold aisle containment are crucial for directing airflow efficiently, preventing the mixing of hot and cold air, and maintaining consistent temperatures throughout server rooms. Proper implementation of these strategies leads to lower Power Usage Effectiveness (PUE) and reduced energy consumption.

Key elements in airflow management include:

- **CRAC/CRAH unit placement**
- **Rack orientation**
- **Use of blanking panels and floor tiles**

- **Proper cable management**

Together, these elements ensure balanced air distribution and help eliminate hot spots, which can compromise equipment performance and reliability.

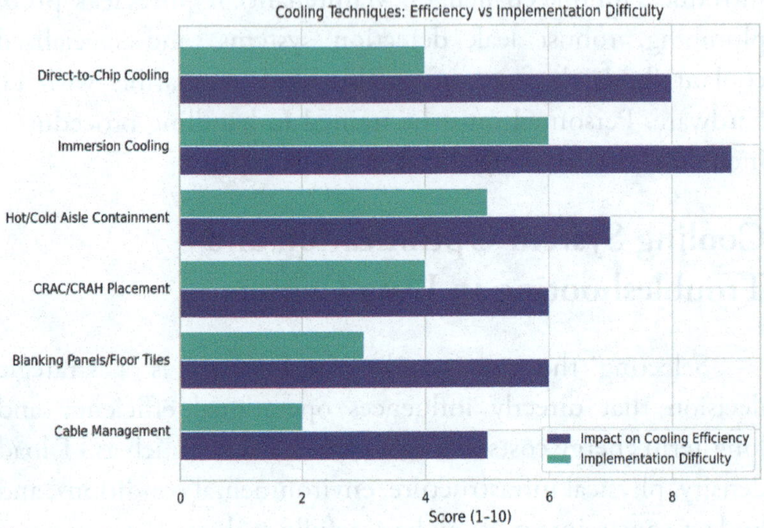

Troubleshooting Cooling Issues

When problems arise, a structured troubleshooting process is critical. Building Management Systems (BMS) provide real time data on multiple environmental and equipment parameters—such as temperature, humidity, pressure, and airflow enabling engineers to isolate the root cause quickly.

For example:

- A sudden temperature spike in a single rack may indicate a CRAC failure, an airflow obstruction, or even an over performing server.

- A rise in chilled water temperature could signal issues with the chiller plant or a drop in pump efficiency.
- Abnormal humidity readings may point to sensor drift, leaking humidifiers, or ventilation issues.

Advanced troubleshooting often involves physical verification in addition to BMS analysis. Engineers may use tools such as thermal cameras, airflow meters, and pressure gauges to confirm BMS indications and localize the issue.

Preventive Cooling System Maintenance

Proactive maintenance is crucial for preventing downtime and extending equipment lifespan. Scheduled activities such as coil cleaning, filter replacement, refrigerant level checks, and fan inspection—must follow the manufacturer's guidelines and be tailored to the system's operational history. A Computerized Maintenance Management System (CMMS) can track maintenance schedules, generate alerts, and log task completion to ensure accountability and consistency.

Preventive maintenance benefits include:

- Higher energy efficiency
- Lower risk of sudden failures
- Improved lifespan of equipment
- Better system reliability during peak demand

Documentation of all inspections and corrective actions not only supports ongoing optimization efforts but also ensures compliance with internal and external standards.

BMS and Real-Time Monitoring for Cooling Systems

The Building Management System serves as the digital control center for modern cooling infrastructure. It aggregates data from an array of sensors and provides engineers with both real time and historical views of the facility's thermal performance.

Key parameters monitored via BMS include:

- Rack and room temperature
- Humidity levels
- Airflow rates through containment systems
- Refrigerant pressure and compressor runtime
- Chilled water flow and temperature

A well-designed BMS dashboard enables engineers to identify trends and deviations quickly. For instance, if a single rack reports elevated temperatures while others remain stable, this could indicate localized airflow restriction, a failing fan, or a faulty temperature sensor. By isolating such patterns, engineers can identify and address the issue before it escalates. Beyond monitoring, the BMS should trigger intelligent alerts based on customized thresholds. These alerts notify staff of potential anomalies such as reduced airflow, chilled water bottlenecks, or unexpected compressor activity so corrective actions can be taken swiftly.

Deeper Insights Through Data Correlation

Effective interpretation of BMS data demands an understanding of the interconnected nature of cooling systems.

For example:

- A rise in compressor runtime may result from a dirty coil, low refrigerant, or motor degradation.
- Reduced airflow in a cold aisle might stem from a dislodged panel, a cable bundle blocking a vent, or even a misconfigured floor tile layout.
- Chilled water supply issues could be caused by scale buildup in heat exchangers, a failing pump, or overloading from IT expansion.

These insights become even more valuable when connected with IT load data. Spikes in power consumption often align with increased thermal output. If cooling systems don't respond properly, it might indicate insufficient capacity or a failure in control logic.

Mastery of cooling system technologies and their integration with BMS platforms is a hallmark of a skilled critical facilities engineer. It requires:

- A deep understanding of system design and airflow strategy
- Hands on experience with troubleshooting and diagnostics
- The discipline to maintain a robust preventive maintenance regimen
- The analytical ability to interpret complex data and anticipate failures

As cooling demands grow with increasing server density and performance requirements, continuous learning and adaptation become essential. By investing in these skills and technologies, data centers can achieve optimal performance, reduce costs, and ensure the uninterrupted operation of mission critical infrastructure.

Cooling System Troubleshooting, Maintenance & Airflow Management

Troubleshooting cooling system issues in a data center begins with a systematic approach grounded in BMS monitoring. Once an anomaly is detected such as a temperature spike, a drop in airflow, or an equipment alarm the engineer must verify the readings using physical sensors or handheld measurement tools. For instance, if a rack shows elevated temperatures, a handheld thermometer can validate whether the sensor is malfunctioning or if there's an actual hot spot. Low airflow should prompt immediate inspection of airflow paths for obstructions, including misplaced cables, blocked vents, or issues with the containment structure.

Checking for refrigerant or chilled water leaks is also critical, requiring the use of specialized detectors or pressure gauges and careful isolation of system components using proper lockout/tagout (LOTO) procedures to ensure safety during repairs.

Addressing airflow issues may involve cleaning or replacing air filters in CRAC or CRAH units, as clogged filters can significantly reduce cooling effectiveness. Verification of fan speed, airflow direction, and motor operation using local controllers or BMS interfaces is essential. If a fan motor fails, prompt replacement is needed to restore airflow. In complex cases, such as refrigerant leaks or mechanical failures, vendor support may be necessary to perform diagnostics and repairs that extend beyond basic in house maintenance.

Proactive maintenance is the foundation of reliable cooling infrastructure. Maintenance schedules must include routine tasks such as coil cleaning, refrigerant checks, and filter replacement, carried out in alignment with equipment manufacturer guidelines and historical performance data. Thorough documentation of all completed maintenance—including the technician's name, date, components serviced, and parts replaced helps in trend analysis and identifying early signs of system degradation. This data driven approach enables targeted replacement of aging components before they cause unexpected failures, significantly reducing unplanned downtime and improving operational efficiency.

In addition to routine maintenance, performance testing plays a crucial role in ensuring system resilience. Regular load testing simulates peak thermal demand, verifying the ability of CRAC, CRAH, and chiller systems to maintain optimal temperatures under stress. Monitoring metrics, such as Power Usage Effectiveness (PUE), provides valuable insights into system efficiency. It helps identify gradual degradation in system performance, whether due to coil fouling, refrigerant inefficiency, or decreased airflow. Early detection through this testing supports proactive intervention and extends the life of equipment.

Effective cooling management also requires close coordination with IT teams. Monitoring power consumption and understanding the IT load profile helps predict heat output and cooling demands. As IT infrastructure expands, cooling capacity must be evaluated and upgraded to avoid bottlenecks. By aligning cooling capabilities with IT growth plans, engineers can ensure stable operating conditions and avoid thermal stress on critical equipment.

A comprehensive emergency response plan is also essential. This plan must outline procedures for cooling system failures including CRAC/CRAH shutdowns, chiller plant malfunctions, and full power outages as well as designated roles, escalation paths, and vendor contact information. Regular simulations and drills ensure personnel are prepared to execute these plans under pressure. An effective response strategy reduces recovery time and limits the impact of failures on business operations and critical IT systems.

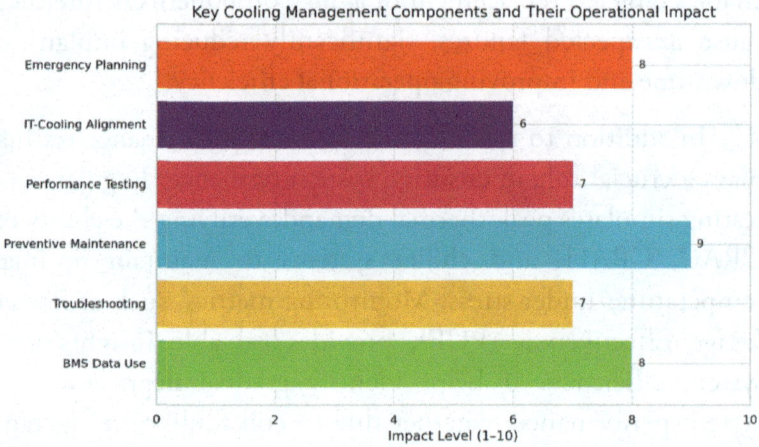

In conclusion, effective cooling system management in data centers relies on a combination of real time monitoring, preventive maintenance, responsive troubleshooting, and robust emergency planning. The BMS is central to this ecosystem, offering visibility into every aspect of cooling performance. When combined with hands on verification, strategic maintenance scheduling, and strong interdepartmental collaboration, this proactive approach supports greater system uptime, lower energy consumption, and enhanced operational reliability.

Airflow Management and Containment Strategies

Efficient airflow management is critical for maintaining ideal environmental conditions within a data center. When airflow is poorly directed, it leads to hot spots, overworked cooling systems, and potential equipment failures. The most effective technique for airflow control is the hot aisle/cold aisle containment strategy, which involves isolating the cold air intake (located at the front of servers) from the hot exhaust (located at the rear). This physical separation prevents the mixing of hot and cold air, significantly improving cooling efficiency.

Cold air is delivered to the cold aisle typically through perforated tiles or directed ducts and drawn through IT equipment where it absorbs heat. The hot exhaust air is confined to the hot aisle and routed back to CRAC/CRAH units or out of the facility. This configuration maintains a stable intake temperature for servers, enabling cooling systems to operate more efficiently, even at higher return air temperatures.

Implementing effective containment involves more than just physical barriers. It requires detailed planning, including the orientation and spacing of server racks, the strategic placement of cooling units, and sealing unused rack spaces with blanking panels. Properly managed airflow contributes to lower PUE, reduces cooling energy costs, and increases overall system stability.

Continuous monitoring ensures the containment strategy remains effective. BMS tools offer insights into airflow volumes, pressure differences, and temperature gradients within the hot and cold aisles. Any deviation from optimal performance such as low airflow readings or rising return temperatures can prompt an investigation into possible blockages, damaged panels, or misconfigured floor layouts.

Advanced Airflow Management and Chiller System Integration

The selection and implementation of containment materials play a pivotal role in determining the effectiveness of a data center's airflow management strategy. Solid panels provide complete separation between hot and cold aisles, offering superior containment and improved cooling efficiency. These are especially effective in high density environments where thermal loads are intense. Perforated panels, while allowing for some mixing of airflows, are often more appropriate in lower density environments where the reduced efficiency is an acceptable tradeoff. Regardless of the choice, airflow capacity must be carefully considered overly restrictive materials can hinder airflow and compromise system performance.

Blanking panels are an essential part of airflow management. These panels fill empty rack spaces to stop cold air from bypassing equipment and escaping into the hot aisle. Without them, the containment system can lose its effectiveness, causing air recirculation, inefficient cooling, and higher energy use. To ensure peak performance, blanking panels must be installed in all unused rack slots and checked regularly for correct placement and physical condition.

Effective airflow management also hinges on the spatial layout of IT equipment and cooling infrastructure. Proper rack spacing prevents bottlenecks and ensures a smooth, unobstructed airflow from the supply to the return paths. The entire data center layout must be designed to support the delivery of cold air to IT inlets and the efficient extraction of hot air exhaust. Computational Fluid Dynamics (CFD) analysis serves as a powerful tool for modeling airflow behavior. It enables engineers to simulate various layouts and identify the optimal arrangement for minimizing hot spots and pressure drops, enhancing overall cooling efficiency.

Monitoring the effectiveness of airflow strategies is equally vital. Key metrics, including temperature, humidity, and airflow velocity, should be monitored at multiple points throughout the facility. These values can be monitored using environmental sensors and visualized through the BMS interface. For example, analyzing temperature differentials between hot and cold aisles gives immediate insight into containment performance. Significant differentials suggest effective separation, while minimal differences often indicate leaks or insufficient airflow. Airflow sensors located within underfloor plenums or ceiling ducts also help detect obstructions and inefficiencies.

When airflow problems arise, troubleshooting begins with a detailed inspection. Leaks around raised floor tiles, damaged blanking panels, or gaps in containment structures are often to blame and must be addressed immediately. If a basic inspection fails to identify the root cause, more advanced tools can be used, such as airflow meters or differential pressure sensors, to pinpoint hidden restrictions. In complex cases, engaging third-party airflow specialists or using smoke visualization techniques can provide a clearer picture of airflow behavior.

Routine maintenance is essential for sustaining optimal airflow performance. This includes cleaning and replacing air filters in CRAC and CRAH units, which, when clogged, severely restrict airflow and reduce cooling effectiveness. Maintaining raised floor systems by inspecting tiles, seals, and cable cutouts prevents uncontrolled airflow leakage and ensures consistent delivery of cold air to server intakes. Maintenance logs should track all interventions, highlighting recurring issues and ensuring accountability.

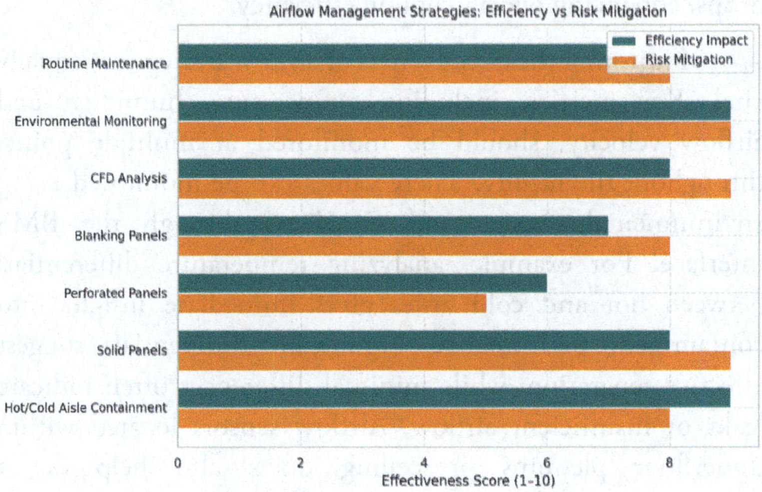

The implications of poor airflow management extend beyond inefficiency. Persistent hot spots can lead to thermal stress on critical hardware, increasing the risk of premature component failure, data corruption, and system downtime. These issues not only affect service reliability but also increase operational costs. Moreover, overcompensating for poor airflow with excessive cooling leads to energy waste and a larger carbon footprint. Thus, effective airflow management is not only a technical concern—it is a business imperative.

To reinforce theoretical knowledge, practical exercises in airflow management are invaluable. These may involve designing containment layouts for various densities, determining optimal CRAC/CRAH placement, and simulating airflow using CFD software. By engaging in hands-on design and evaluation scenarios, engineers can refine their understanding of airflow behavior and improve decision-making skills. These exercises support a data driven, proactive approach to airflow optimization.

Integral to this cooling ecosystem are chiller systems, which serve as the backbone of large scale data center cooling. Chillers provide chilled water to CRAH units, which in turn cool the air circulated across server racks. These systems typically consist of key components such as compressors, evaporators, condensers, and expansion valves—each vital for ensuring efficient thermal transfer. The operation begins with warm return water from CRAH units entering the evaporator. Here, the refrigerant absorbs heat from the water, which is then cooled and sent back to the air handling units. The heated refrigerant is compressed, routed to the condenser, and cooled using either air or water before the cycle is restarted.

Maintaining chiller performance is paramount. Routine tasks include monitoring refrigerant levels, inspecting water flow rates, testing pressure sensors, and cleaning heat exchanger surfaces. Scaling or fouling within the system can reduce heat transfer efficiency, while flow imbalances can lead to uneven cooling across the facility. Therefore, regular performance assessments and vibration analysis of mechanical components are recommended to catch issues before they escalate into system failures.

The critical facilities engineer must also understand how chiller operation aligns with broader facility metrics, such as Power Usage Effectiveness (PUE) and water consumption. This means optimizing setpoints based on seasonal conditions, implementing variable speed drives on pumps and compressors, and coordinating with BMS data to schedule chiller operation efficiently. With advanced DCIM and BMS tools, engineers can track chilled water supply and return temperatures in real time, adjusting operational parameters to reduce energy consumption without sacrificing performance.

In summary, the effective integration of airflow management and chiller systems is crucial for maintaining a high performance data center. By understanding the nuances of containment strategies, airflow optimization, and chiller operation, critical facilities engineers can create environments that not only maintain IT reliability but also support long term operational sustainability. As cooling technologies evolve, staying informed and hands on remains key to excellence in data center thermal management.

Airflow Management and Chiller Systems in Data Centers

Effective airflow management is a cornerstone of efficient data center operations, directly influencing cooling performance, equipment longevity, and energy consumption. Without proper containment and directional control, airflow inefficiencies can lead to hot spots, elevated operating temperatures, and potential hardware failure. The most widely adopted strategy is hot aisle/cold aisle containment, which physically separates incoming cool air from outgoing hot exhaust, thereby reducing mixing and maximizing thermal efficiency. In this configuration, cold air is directed to the front of the server racks, while hot air is extracted from the rear, typically using solid or perforated panels, depending on the cooling needs and rack density. Solid panels provide better isolation in high density environments, while perforated panels offer flexibility in less demanding scenarios. Blanking panels are critical for sealing empty rack spaces and preventing recirculation, ensuring that cold air is used efficiently and not wasted.

The design of airflow pathways requires more than simple physical separation.

Equipment layout, rack spacing, and tile placement all play essential roles in optimizing air movement. Improper spacing can lead to bottlenecks or uneven distribution, which affects the performance of cooling units such as CRAC (Computer Room Air Conditioning) and CRAH (Computer Room Air Handling) systems. Computational Fluid Dynamics (CFD) modeling is frequently employed during design and retrofitting to simulate airflow and identify inefficiencies before physical deployment. Post deployment, the effectiveness of airflow strategies should be monitored using environmental sensors and Building Management System (BMS) data. Monitoring parameters such as cold aisle and hot aisle temperature differentials provides valuable feedback on the performance of containment systems. Narrow temperature gaps may indicate air leakage or improper sealing, while broader gaps suggest successful separation and optimal heat removal.

When airflow anomalies are detected, troubleshooting begins with a thorough inspection of the containment infrastructure. Leaks in raised floor tiles, misaligned blanking panels, or damaged panels can all compromise airflow patterns. In such cases, correcting physical issues often restores thermal balance. When issues persist, tools like pressure differential meters and thermal cameras can help identify hidden causes.

Air filters in CRAC/CRAH units also require attention; clogged filters reduce airflow, increase energy consumption, and lower cooling capacity. Regular cleaning or replacement, as per the manufacturer's guidelines, is essential. Maintaining raised floors also contributes to airflow reliability damaged tiles or broken gaskets can allow air to escape, undermining containment effectiveness.

Beyond airflow, understanding chiller systems is fundamental to managing large scale cooling infrastructure. Data centers typically employ two main types: absorption chillers and centrifugal chillers. Absorption chillers utilize heat often sourced from steam or waste heat sources to drive a thermodynamic cycle involving a refrigerant and an absorbent, typically lithium bromide. The process includes the generator, absorber, evaporator, and condenser, forming a continuous loop that produces chilled water. These chillers are highly efficient in environments where waste heat is plentiful. However, reduced heat input can cause a decline in performance, making it essential to closely monitor the heat source and flow.

Centrifugal chillers, on the other hand, rely on mechanical compressors to cycle refrigerant through a high efficiency loop comprising a compressor, condenser, expansion valve, and evaporator. These systems are ideal for large cooling loads, offering excellent performance at scale. Key performance indicators such as pressure levels, temperature differentials, and compressor vibration should be routinely tracked.

Issues like refrigerant leaks or rising condenser temperatures often show up as deviations in these readings. Proper diagnostics require a thorough understanding of the entire refrigerant cycle and the ability to connect sensor data with mechanical performance. Preventive checks such as monitoring operating currents and assessing compressor integrity are key to avoiding catastrophic failures.

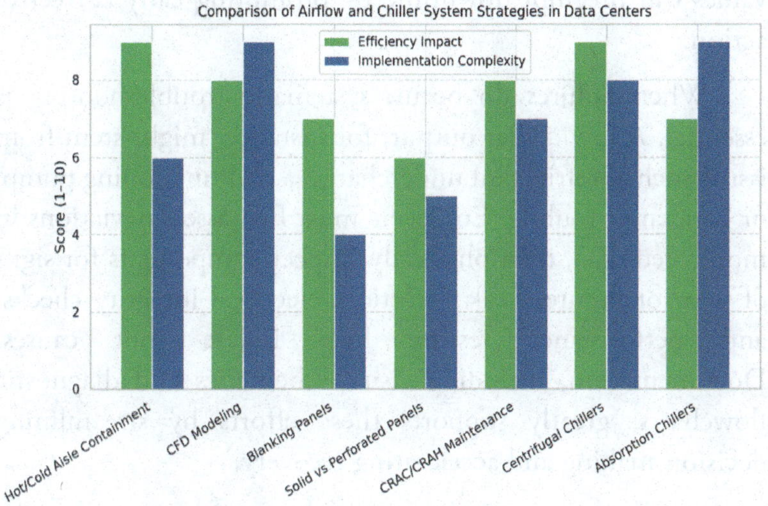

A critical supporting element of any chiller system is water treatment. Untreated or poorly treated water leads to scale buildup and corrosion, which reduces thermal transfer and damages internal components. Chemical dosing to soften water and inhibit corrosion must be precisely controlled and continuously monitored through water sampling and lab analysis. Neglecting this aspect not only shortens the lifespan of the chillers but also threatens the associated infrastructure, including pumps, valves, and pipework.

Comprehensive monitoring is essential for the operational success of both airflow and chiller systems. Modern chillers are integrated with advanced control systems that provide real-time metrics on flow rates, energy use, and system performance. These insights enable engineers to identify trends, predict failures, and initiate timely maintenance. In tandem, the BMS aggregates data from across the facility—airflow, temperatures, pressure points enabling a holistic view of cooling performance. Alerts triggered by abnormal values can preempt downtime by prompting early corrective action.

When failures do occur, systematic troubleshooting is essential. A low chiller output, for instance, might stem from issues such as refrigerant undercharge, a malfunctioning pump, or condenser fouling. Engineers must first assess deviations in monitored data, then physically inspect components for signs of wear or failure. Leak detection, electrical integrity checks, and performance testing help isolate root causes. Documentation, including visual schematics and diagnostic flowcharts, greatly supports these efforts by streamlining decision making and accelerating recovery.

In conclusion, efficient data center cooling hinges on the integration of advanced airflow management and robust chiller infrastructure. Mastery over both domains, along with real-time monitoring and preventive maintenance, enables critical facilities engineers to deliver maximum uptime, energy efficiency, and hardware longevity.

As cooling demands increase alongside rising rack densities, a proactive, systems based approach will remain essential in ensuring that modern data centers operate smoothly and sustainably.

Chiller and HVAC Preventative Maintenance in Data Centers

Regular maintenance is foundational to sustaining the performance and reliability of chiller systems within a data center. These systems, often running continuously under high demand, require scheduled inspections, cleaning, and component servicing to operate effectively. A robust maintenance program includes routine tasks such as filter replacement, lubrication of moving parts, visual inspections, and functional testing of key components like pumps, fans, and valves. To ensure consistency and accountability, a detailed maintenance schedule should be established and rigorously adhered to.

This schedule must outline the nature of each task, its required frequency, necessary safety protocols, and specific procedural steps. Only qualified personnel, trained in both system operation and workplace safety, should carry out these tasks. Thorough documentation of each maintenance activity including findings, parts replaced, and any abnormalities serves as both a historical reference and a predictive tool, enabling data center teams to recognize performance trends and intervene early. Regular maintenance not only extends the lifespan of the chiller system but also enhances energy efficiency, minimizes the risk of system failure, and reduces the financial impact of unplanned downtime.

More broadly, preventative maintenance is a strategic necessity for all HVAC systems operating in data centers. Failure to implement such programs can lead to cascading failures, data center outages, and unnecessary capital expenditures. A comprehensive preventative maintenance plan begins with a tailored schedule that considers equipment types, environmental factors, and manufacturer recommendations. For instance, a facility in a humid region may require more frequent air filter replacements than one in a temperate climate due to increased particulate buildup.

This schedule should encompass various maintenance activities, including visual inspections, airflow checks, coil cleaning, and electrical connection verifications. By recording each completed task and linking these records with system performance, engineers can identify emerging issues and adjust the maintenance schedule proactively. This organized data collection also allows for trend analysis, which can reveal subtle changes in system behavior that may indicate upcoming failures.

Visual inspections are a cornerstone of this preventative approach. Conducted at regular intervals, these inspections involve examining all HVAC components for early signs of wear, leaks, damage, or unusual behavior. Specific focus areas include coil corrosion, belt integrity, pulley alignment, electrical wiring condition, and the cleanliness of moving parts, such as fans. Engineers should thoroughly document inspection results, supplementing written notes with high-resolution photographs of the affected components. This visual evidence is invaluable for diagnosing recurrent issues and verifying equipment condition over time.

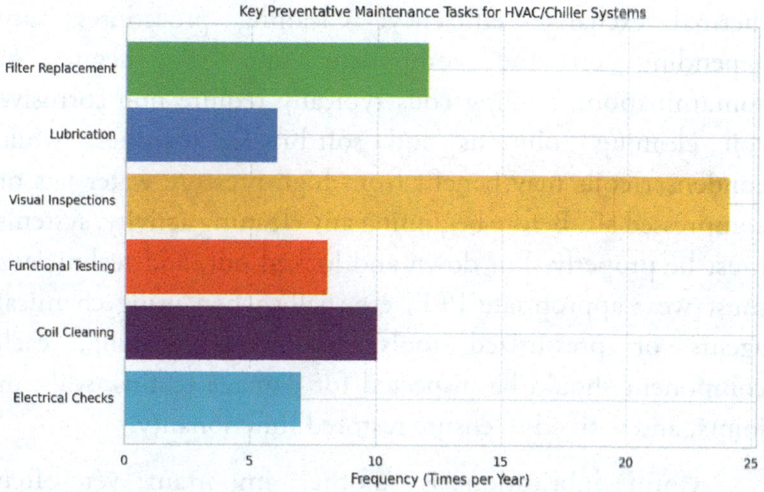

Air filters, while often overlooked, are critical to maintaining optimal system performance. These filters act as the first line of defense against airborne contaminants that can reduce airflow and efficiency. The frequency of filter replacement depends on factors such as environmental cleanliness, the type of filter used, and the operational hours of the HVAC units.

Regardless of these variables, all filter replacements must begin with strict adherence to lockout/tagout (LOTO) protocols to prevent accidental startup during servicing. Once the system is safely de energized, technicians can access the filter housing, which is usually designed for quick access, and replace the filters. Used filters should be disposed of in accordance with local environmental regulations to avoid contamination or regulatory violations.

Another vital component of HVAC maintenance is cleaning of key system elements such as cooling coils, condenser coils, and fan assemblies. Over time, these parts accumulate dust and debris that restrict airflow and reduce thermal exchange efficiency. Cleaning procedures vary depending on the component and the degree of contamination. Cooling coils typically require non corrosive coil cleaning solutions and soft bristled brushes, while condenser coils may benefit from high-pressure water jets or compressed air. Before beginning any cleaning activity, systems must be properly shut down and locked out, and technicians must wear appropriate PPE, especially when using chemical agents or pressurized tools. Following cleaning, each component should be inspected for damage to fins, seals, or joints, and verified to ensure restored functionality.

Motor lubrication is another important yet often overlooked part of preventive maintenance. HVAC motors, especially those powering fans and pumps, wear out more quickly because of continuous use. Regular lubrication reduces friction, prevents overheating, and helps extend motor life. Technicians need to follow manufacturer specific instructions for the right type of lubricant and proper intervals. Over lubricating or using incompatible products can damage motor parts, just as neglect can cause bearing failure and energy waste. As always, proper lockout/tagout procedures must be followed before any lubrication or mechanical work.

In closing, chiller systems and HVAC infrastructure require not only technical understanding but also disciplined execution of maintenance protocols. Establishing and following a comprehensive preventative maintenance schedule ensures optimal performance, reduces the likelihood of critical failures, and helps maintain a safe operating environment. The collection and analysis of maintenance data, combined with visual inspections and equipment testing, enable data center engineers to adopt a proactive approach. Investments in training, documentation, and safety compliance ultimately translate into fewer disruptions, lower costs, and a more resilient and efficient facility.

Preventative maintenance extends beyond routine tasks, such as filter replacement and cleaning. It also includes inspecting and maintaining other components within the HVAC system. This may involve checking refrigerant levels in refrigeration systems, inspecting belts and pulleys for wear, verifying the proper operation of sensors and controllers, and testing safety mechanisms, such as emergency shut off switches.

Regular monitoring of system parameters, such as temperature and pressure, using the Building Management System (BMS), provides valuable insights into system performance and helps in the early detection of potential problems. Trends detected within the BMS data can indicate upcoming failures or necessary adjustments to the maintenance schedule. This proactive monitoring significantly reduces the likelihood of unexpected breakdowns, allowing for timely intervention.

The effectiveness of a preventative maintenance program relies heavily on meticulous record keeping. A well-maintained log detailing all performed maintenance activities, including the date, time, specific tasks performed, parts replaced, and any observations made, is crucial for effective tracking and analysis. This detailed record helps identify trends, anticipate potential issues, and demonstrate compliance with industry standards and regulations.

The data contained within the log can also be used to optimize the maintenance schedule, ensuring that resources are used effectively and that the program remains proactive. Maintaining a comprehensive record also supports the justification for maintenance expenditures and helps demonstrate a commitment to efficient and reliable data center operations. By implementing a well defined schedule, performing regular inspections, diligently carrying out maintenance tasks, and maintaining meticulous records, data center engineers can significantly reduce downtime, extend the lifespan of equipment, and optimize overall system performance.

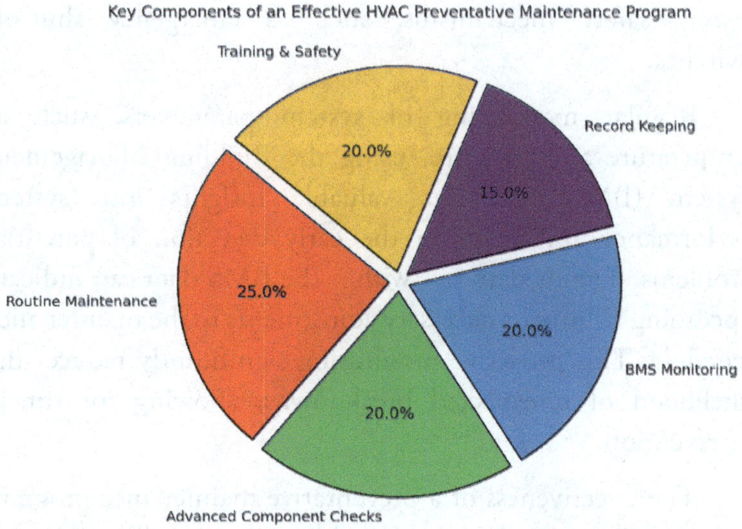

This proactive approach not only safeguards the integrity of critical IT infrastructure but also contributes to significant long term cost savings. Investing in training and implementing a robust maintenance program ultimately translates to a more reliable, efficient, and cost effective data center operation. Remember that safety is paramount throughout all maintenance procedures. Always follow appropriate lockout/tagout procedures, use necessary PPE, and adhere to the manufacturer's instructions to ensure the safety of personnel and the integrity of the equipment.

Chapter 3: UPS Systems and Power Distribution

Understanding Uninterruptible Power Supply (UPS) Systems

Mastery of Uninterruptible Power Supply (UPS) systems is crucial for any critical facilities engineer, as these systems form the backbone of data center power continuity and reliability. During utility failures or fluctuations, UPS systems provide immediate backup, safeguarding sensitive IT equipment from damage and ensuring continuous operation. A strong understanding of UPS architectures online, offline, and line-interactive and their operational principles enables engineers to make informed decisions tailored to specific data center needs.

UPS System Classifications and Architectures

The most robust of these systems is the online UPS, often regarded as the industry benchmark for power protection. This system operates by converting incoming AC power to DC via a rectifier, storing it in a battery, and then converting it back to AC through an inverter before delivering it to the load. This continuous double conversion process ensures that connected devices receive clean, isolated power, free from voltage spikes, brownouts, or frequency variations. In the event of a utility power failure, there is no transfer delay the battery seamlessly supplies energy to the inverter, ensuring uninterrupted power to the load. Online UPS systems are ideal for mission critical environments, including large scale data centers, hospitals, and financial institutions, where even momentary power fluctuations are unacceptable. Their modular design supports scalability and enables hot swappable maintenance, reducing system downtime and allowing for power capacity expansion in line with growing infrastructure needs. Despite their higher initial cost and greater energy consumption, modern online UPS units incorporate energy-efficient technologies that reduce operating costs and environmental impact.

In contrast, offline (standby) UPS systems provide a simpler and more economical approach. Under normal conditions, power flows directly from the utility to the equipment, bypassing the inverter. The system monitors the input line and, upon detecting a fault or blackout, activates the battery and inverter to supply power to the load.

Although this switchover typically occurs within a few milliseconds, that short delay can disrupt sensitive electronics or lead to data loss in certain applications. For this reason, offline systems are best suited for environments where equipment is more tolerant of power interruptions, such as small offices, home workstations, or basic point sale systems. Their straightforward architecture makes them easy to deploy and maintain, and they deliver significant energy efficiency when power quality is stable.

Falling between these two extremes is the line interactive UPS, which offers a middle ground in terms of cost, complexity, and protection. This system features a built in voltage regulator that actively compensates for minor sags and surges, eliminating the need to switch to battery power. It utilizes a transformer and an inverter, which remain in standby mode until significant disturbances occur. When necessary, the system quickly switches to battery mode with a typically shorter delay than offline models.

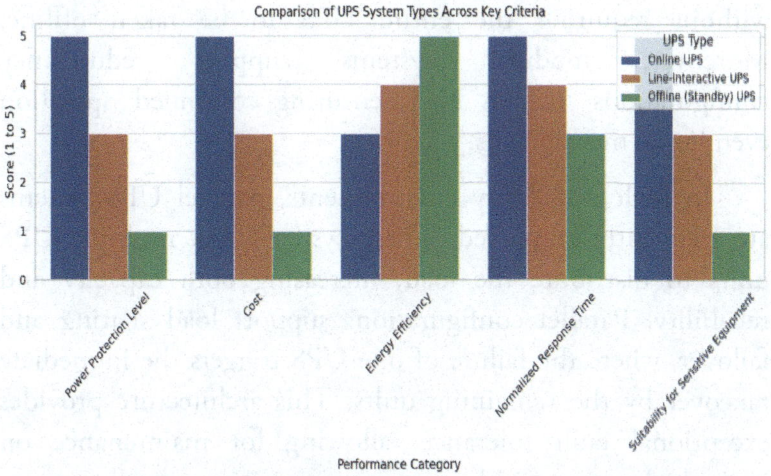

This makes line interactive systems suitable for small to mid size server rooms or network closets, where power quality may be inconsistent but continuous uptime is not as critical as in enterprise data centers. These systems offer a compelling balance higher protection than offline UPS at a cost considerably lower than full online systems, making them ideal for moderately sensitive applications and budget-conscious environments.

Scalability, Redundancy, and Advanced Topologies

Beyond functional classification, UPS systems differ significantly in terms of capacity and configurability. Larger facilities often adopt modular UPS architectures, which feature multiple interchangeable power modules within a single frame. These allow for seamless scalability additional modules can be inserted as the power demand grows, without requiring the replacement of the entire system. This approach offers enhanced serviceability, as failed modules can be replaced without requiring the entire UPS to be taken offline. Moreover, modular systems support redundancy configurations such as N+1, ensuring continued operation even if one module fails.

In high availability environments, parallel UPS systems are frequently employed. These systems link multiple UPS units to distribute the load, increasing both capacity and reliability. Parallel configurations support load sharing and failover, where the failure of one UPS triggers the immediate takeover by the remaining units. This architecture provides exceptional fault tolerance, allowing for maintenance on individual units without compromising overall system availability. The scalability and resilience of parallel systems make them well suited for enterprise data centers, financial trading platforms, and healthcare facilities that demand uninterrupted power under all conditions.

UPS Topology	Key Features	Scalability	Redundancy Options	Ideal Use Cases
Modular UPS	Interchangeable power modules, hot-swappable, compact footprint	High – add modules as needed	N+1, N+2	Growing data centers, enterprise IT, and co-location
Parallel UPS	Multiple full UPS units,	High – add	N+1, 2N, 2(N+1)	Mission-critical

	load sharing, high fault tolerance	UPS units		systems, financial, and healthcare
Standal one UPS	Fixed capacity, traditional single-unit design	Low – must replace to expand	Limited or none	Small/medi um offices, branch locations

Operational Considerations and Deployment Strategy

Selecting the appropriate UPS system involves evaluating not only power requirements but also risk tolerance, budget, and long term growth plans. Online UPS systems offer the highest level of protection but are more expensive and complex to operate. Offline systems are cost effective but suited only for non critical loads. Line interactive systems present a practical compromise. Modular and parallel designs further enable data centers to adapt dynamically to changing loads and service demands.

UPS systems must be deployed in conjunction with battery monitoring, load assessments, and integration with Building Management Systems (BMS) to ensure real time insight into system status. Preventative maintenance, such as regular battery testing and thermal scanning, is essential for prolonging the lifespan and avoiding catastrophic failures.

UPS System Components, Monitoring, and Troubleshooting

A fully functional Uninterruptible Power Supply (UPS) system relies on the seamless interplay of its core components, each of which fulfills a critical role in ensuring continuous power delivery. At the front end, the rectifier converts incoming alternating current (AC) from the utility grid into direct current (DC), which charges the battery bank and supplies power to the inverter. During a power disturbance or outage, the inverter draws from the battery to convert DC back into clean AC power for the connected load, ensuring uninterrupted operation of sensitive IT equipment.

The control unit acts as the brain of the UPS, continuously monitoring system parameters input voltage, battery health, and output power and managing transitions between power sources. It ensures smooth switching between mains and backup power, responding instantly to voltage fluctuations or outages. Additionally, bypass switches allow power to be rerouted around the UPS for maintenance or in the event of a system failure. These must be operated with extreme caution and in accordance with strict safety protocols, as they often involve live electrical circuits. The strategic integration of these components forms a resilient power delivery system capable of protecting critical infrastructure under a wide range of conditions.

Selecting the right UPS system requires more than choosing a topology—it demands careful attention to sizing, integration, and ongoing maintenance. Capacity planning is essential; the system must not only meet current IT load requirements but also account for potential future growth.

Proper integration into the overall power distribution network ensures load balancing and redundancy, which are essential for avoiding single points of failure. Maintenance routines, including scheduled battery replacements and firmware updates, must be consistently followed to sustain long-term reliability. Failure to adhere to these best practices increases the risk of unplanned outages and escalates operational costs due to equipment failure or service disruptions.

The decision to choose a specific UPS architecture whether offline, line-interactive, or modular online must align with the facility's unique needs. Factors such as risk tolerance, load sensitivity, budget, and expected power stability all influence the choice. Smaller setups might find sufficient protection with line-interactive or offline systems, while larger enterprise environments require modular or parallel online systems to support high availability operations. Regardless of the configuration, a UPS is a strategic investment. A well informed, data driven selection ensures resilience, scalability, and long term performance that align with organizational goals.

Monitoring and Maintaining UPS Systems

Reliable UPS operation hinges not only on its design but also on active monitoring and thorough maintenance. Effective UPS monitoring includes real time data collection, preventative upkeep, and targeted troubleshooting. Modern UPS systems come with advanced diagnostics and alarm features that offer immediate insight into system health. These capabilities allow engineers to identify emerging issues early, respond quickly, and prevent extended outages.

Alarms serve as the first line of defense in detecting system anomalies. These can range from non critical warnings, such as temperature fluctuations or fan failures, to critical alerts, like inverter failure or battery discharge faults. Accurate interpretation of these alarms depends on understanding system documentation and error codes, which detail the nature, urgency, and necessary responses for each issue. For example, a low battery voltage alarm might indicate aging batteries or charging problems, prompting inspection and potential replacement. Conversely, an inverter fault requires immediate action, such as shifting the load to bypass mode or activating redundant systems.

Battery maintenance is perhaps the most critical aspect of UPS reliability. Traditionally, lead acid batteries have been used; however, lithium ion technologies are gaining traction due to their longer lifespan, reduced maintenance needs, and improved performance under different temperature conditions. Battery health is assessed through voltage checks, impedance tests, and full load testing, with the latter providing the most accurate measure of a battery's capacity in real world conditions. Routine battery testing intervals are based on manufacturer recommendations and operational factors like ambient temperature, age, and load characteristics. Skipping these tests can hide vulnerabilities until a failure occurs.

Testing and maintenance activities must adhere to strict safety protocols. UPS systems operate at high voltages, and improper handling can result in serious injury or damage to the equipment. Lockout/tagout (LOTO) procedures be followed during maintenance, and technicians should use personal protective equipment (PPE), insulated tools, and appropriate signage to maintain a safe work environment. Visual inspections for corrosion, swelling, or leaks complement electrical testing and help identify batteries approaching the end of their life before they compromise system reliability.

Remote monitoring tools are playing an increasingly vital role by enabling engineers to access UPS performance data and alerts from any location. These tools integrate with Building Management Systems (BMS) and offer dashboards displaying power quality metrics, historical trends, and real time fault data. Trend analysis allows engineers to correlate battery degradation with environmental conditions or load changes, providing predictive insights that support better maintenance planning.

In summary, a UPS system is far more than a backup power source it is a dynamic, multi-component infrastructure that safeguards the heart of data center operations. Understanding the role and function of each component, ensuring proper sizing and integration, and committing to robust monitoring and maintenance practices form the foundation of UPS reliability.

By embracing proactive strategies and continuous learning, critical facilities engineers ensure that power disruptions—no matter how sudden do not compromise mission critical IT functions. As UPS technologies evolve, staying current with new architectures, battery chemistries, and diagnostic tools remains essential for maintaining data center resilience and operational excellence.

Load Balancing, Troubleshooting, and Generator Systems in Critical Power Infrastructure

Maintaining proper load balance is critical for the optimal performance and long-term reliability of Uninterruptible Power Supply (UPS) systems. The UPS must be appropriately sized to support all connected loads, with an adequate safety margin to mitigate the risk of overloading. Continuous monitoring of real time load levels typically expressed as a percentage of total capacity is essential to ensure the system remains within its operational thresholds. Consistently operating below the maximum load capacity promotes energy efficiency, extends battery life, and minimizes the risk of unexpected shutdowns.

Overloads not only accelerate battery degradation but can also cause thermal stress or complete system failure. Through regular analysis of load data, operators can proactively identify risks and implement mitigation strategies such as load shedding, which prioritizes critical systems temporarily disconnects non essential loads during peak demand. This ensures the continuity of operations without compromising essential functions.

Troubleshooting UPS systems requires a methodical process, beginning with a comprehensive review of system alerts, load behavior, and input/output parameters. For example, a low battery voltage alarm may suggest undercharging or battery wear. Diagnosis should begin with verifying the rectifier output, as insufficient voltage or current from the charging system can prevent batteries from reaching full capacity. If the rectifier is functioning properly, the next step is to assess battery health, checking age, impedance values, and past load test results to determine whether replacement is needed. In cases where the UPS indicates an overload, engineers should analyze the load distribution and consider shedding non critical loads or upgrading the UPS to accommodate growing demand. An inverter failure a critical fault—typically requires professional intervention. Depending on the system's design, this could involve replacing an inverter module or bypassing the system until repairs are completed.

Accurate interpretation of UPS system data is fundamental to effective fault isolation and sustained uptime. Modern UPS units produce a comprehensive set of operational metrics, including input/output voltage, current, frequency, battery temperature, power factor, and system efficiency. These parameters act as diagnostic indicators: for instance, a dip in input voltage may signal grid instability, whereas persistently high output current could indicate circuit overloading. Low battery voltage may result from sulfation, deep discharge cycles, or cell failure. Proficiency in correlating these metrics allows engineers to localize faults, validate system behavior, and take prompt corrective action. Many contemporary UPS systems support remote monitoring, enabling real time data collection via centralized dashboards or mobile platforms. This functionality is particularly advantageous in distributed or multi site operations, facilitating rapid responses to system alerts or environmental anomalies without requiring on site presence.

Preventative maintenance remains the cornerstone of UPS reliability. Scheduled inspections, cleaning, and battery testing are essential for preventing minor issues from escalating into critical failures. During maintenance, trained personnel should inspect for loose terminals, signs of corrosion, heat damage, and check airflow to prevent overheating.

Accumulated dust and debris, especially near ventilation inlets and electronic components, must be regularly cleared to maintain proper cooling and system integrity. Detailed maintenance records are critical, offering insight into component performance, replacement schedules, and health trends over time. This historical data not only aids in troubleshooting but also supports compliance and informed upgrade planning. Following manufacturer maintenance guidelines ensures that system performance remains aligned with design specifications and service life expectations.

In conclusion, maintaining UPS systems involves a coordinated strategy of load monitoring, systematic diagnostics, and proactive maintenance. Engineers must interpret real-time system data effectively, respond swiftly to alarms, and implement load management strategies to avoid system stress. With advancements in remote monitoring and predictive analytics, today's UPS systems offer more control and reliability than ever before. By investing in robust operational practices and staying abreast of technology developments, data centers can ensure power continuity and optimize their infrastructure for long term success.

Generator Systems in Data Center Power Architecture

While UPS systems offer short term power backup, generators are essential for long term backup during extended utility outages. They ensure critical IT operations continue without disruption, often bridging the gap for hours or even days, depending on fuel availability and load requirements. Understanding generator technology and operational protocols is fundamental to designing a resilient data center infrastructure.

The most common generator in data centers is the diesel generator, known for its reliability, high power output, and durability. However, natural gas generators are becoming more popular due to lower emissions, longer fuel availability through pipeline infrastructure, and cheaper operating costs in some areas. The choice between diesel and natural gas systems depends on several factors: fuel availability, environmental laws, space limitations, and overall cost of ownership. For example, in remote locations without gas lines, diesel is usually preferred, while facilities in urban areas with strict emission rules might choose gas powered units.

Proper sizing of generator systems is crucial. The generator must not only accommodate the full IT and HVAC load but also provide headroom for future expansion and handling inrush currents during startup. This is typically achieved through comprehensive load profiling, factoring in peak demand, redundancy requirements, and start up sequencing of high power equipment. An undersized generator can fail during peak load conditions, while oversizing can lead to inefficient fuel use and maintenance challenges.

Pre operational checks are mandatory before starting a generator. These include:

- Verifying fuel levels and fuel quality to support the intended runtime.
- Inspecting the engine compartment for leaks, wear, or foreign objects.
- Checking oil and coolant levels, ensuring they meet operational thresholds.
- Confirming battery voltage for starter systems and auxiliary power needs.
- Ensuring the exhaust system is intact and unobstructed to prevent hazardous gas accumulation.

All findings must be documented, and any anomalies corrected before starting the generator. Keeping detailed checklists for these inspections ensures consistency, safety, and regulatory compliance. Documentation also helps with trend analysis, enabling facility managers to anticipate maintenance needs or systemic issues before they become critical.

Generator Systems and Transition to PDUs in Data Center Power Management

Starting a generator in a data center environment demands strict compliance with standardized safety protocols, particularly **Lockout/Tagout** (LOTO) procedures. These ensure the generator is safely isolated from the main power grid, preventing accidental energization during maintenance or manual startup. While most modern generators employ automated start sequences, a solid understanding of the manual startup process remains critical for effective troubleshooting. The typical startup procedure involves activating the main circuit breaker, engaging the starter system, and closely observing system behavior as it comes online. During operation, key performance metrics such as engine speed, oil pressure, coolant temperature, and voltage output must be closely monitored. Deviations from established thresholds can signal emerging problems that require immediate attention. Advanced control panels or digital monitoring interfaces often streamline this process, offering real time feedback and alerts for prompt corrective action.

A fundamental aspect of generator reliability is **fuel system management**. Diesel generators require clean, high-grade fuel to sustain optimal engine performance and prevent injector clogging or filter obstructions. Fuel storage tanks must be adequately sized to support extended runtimes during prolonged outages and should be routinely inspected for leaks, microbial growth, or sediment accumulation. The implementation of fuel polishing and filtration systems helps maintain fuel purity by removing water and contaminants. For natural gas generators, maintaining consistent gas pressure and clean supply lines is equally critical, as supply inconsistencies can negatively affect performance. Both fuel types necessitate regular system diagnostics, and any anomalies such as pressure loss, corrosion, or filtration blockages must be addressed immediately to ensure operational readiness.

Maintenance is the cornerstone of generator longevity and performance. Following the manufacturer's maintenance schedule, tasks such as oil changes, filter replacements, coolant checks, and battery inspections should be performed at specified intervals.

These activities not only optimize system efficiency but also mitigate the risk of failure during critical operations. All maintenance actions should be thoroughly documented, including inspection dates, findings, and corrective measures. This documentation is valuable for trend analysis, warranty validation, regulatory compliance, and long term infrastructure planning. Environmental factors such as temperature fluctuations, humidity levels, and airborne particulates also influence maintenance frequency and must be incorporated into maintenance strategies. Neglecting timely service jeopardizes system integrity and increases the risk of unplanned downtime and costly repairs.

To validate a generator's performance, load testing is essential. These tests simulate real operating conditions by placing the generator under controlled electrical loads equivalent to the data center's peak demand. This ensures that the generator can reliably support critical systems during an outage. Load tests help identify weaknesses in fuel delivery, voltage regulation, and thermal management. Exercise runs, typically conducted without full load, are also important for ensuring reliable starts and verifying the correct function of the Automatic Transfer Switch (ATS). Both load tests and exercise runs should be scheduled regularly, and all outcomes—especially anomalies such as undervoltage, frequency drift, or extended start times must be recorded and addressed. Regular testing ensures that the system functions not just in theory, but under real world demand, with rapid transition times and stable output.

The transition from utility power to generator power is managed through the ATS, which automatically senses a power failure and initiates the switch to generator output. Critical facilities engineers must monitor this process, ensuring that voltage, frequency, and phase synchronization are within tolerance before the load is transferred. In some situations, manual override or troubleshooting may be required, particularly if the ATS fails to respond or encounters faults. Engineers must have a deep understanding of ATS logic, be prepared to execute manual overrides, and follow predefined escalation protocols to ensure uninterrupted system functionality. Timely, accurate responses to transfer anomalies are essential to maintaining **Service Level Agreements (SLAs)** and upholding business continuity. Regular training sessions and simulated failure drills enhance preparedness and increase operator confidence during actual emergencies.

Beyond daily operation, critical facilities engineers play a pivotal role in vendor coordination, SLA enforcement, and budget planning for generator systems. They must interpret maintenance reports, verify task completion, and justify capital or operational expenditures for repairs, fuel supply, and system upgrades. Their in depth knowledge enables them to align generator capacity with data center growth while minimizing operational risks. By integrating technical expertise, strategic planning, and rigorous documentation, engineers ensure that backup power systems function as a reliable safety net rather than a point of failure. Proactivity, not reactivity, defines success in generator system management anticipating problems before they occur is far more effective than reacting to emergencies after they occur.

Once backup power systems—such as UPS units and generators are secured, operational focus shifts to **Power Distribution Units (PDUs)**, which form the critical interface between centralized power infrastructure and the IT equipment housed within individual server racks. PDUs enable precise power allocation, offering granular monitoring of current, voltage, and energy consumption at the device or outlet level. Understanding their deployment, types, and monitoring capabilities is essential for efficient capacity planning, fault detection, and load balancing across the data center environment.

Generator Systems and PDU Transition Overview

Topic	Description
Generator Startup Procedures	Follow safety protocols (LOTO) and understand both manual and automated startup procedures. Monitor the main breaker, starter, and online transition.
Key Performance Monitoring	Monitor engine speed, oil pressure, coolant temperature, and voltage output using control panels or digital interfaces.
Fuel Management	Ensure clean, quality fuel (diesel or gas), check for leaks, and use filtration systems. Regular inspections are critical.

Preventive Maintenance	Follow the manufacturer's maintenance schedule (oil, filters, coolant, battery), document all findings, and factor in environmental conditions.
Load Testing	Simulate real loads to test generator capacity, identify weak spots (fuel, voltage, thermal), and confirm ATS function.
Automatic Transfer Switch (ATS)	ATS detects a power failure, switches to the generator, and ensures voltage and frequency synchronization. Manual override may be required.
Critical Facilities Engineer Role	Oversee vendor tasks, enforce SLAs, manage budgets, align capacity with data center growth, and ensure documentation.
Transition to PDUs	PDUs distribute power to IT racks, enable granular monitoring (current, voltage), facilitate fault detection, and aid in load balancing.

Power Distribution Units (PDUs)

Power Distribution Units (PDUs) are essential components in modern data center architecture, bridging the gap between centralized power systems such as UPS units and generators and individual IT hardware. More than simple power strips, today's PDUs are intelligent devices that support granular monitoring, remote management, and power optimization. Their correct implementation directly affects operational efficiency, system uptime, and the longevity of connected equipment.

One of the primary functions of PDUs is load balancing, which is especially vital in environments with dynamic power demands. By evenly distributing electrical loads across multiple PDUs, engineers prevent overloading individual circuits, which can lead to breaker trips or equipment shutdowns. For instance, if a single PDU becomes heavily loaded and experiences a power surge, it could trip its circuit breaker, triggering a localized outage affecting multiple servers. Through strategic distribution of the load, such risks are minimized, ensuring resilience and consistent uptime across the data center. Load balancing also improves energy utilization, allowing the facility to operate more efficiently while minimizing waste and costs.

Selecting the appropriate PDU type for a specific deployment depends on various criteria, including power capacity requirements, monitoring needs, and rack layout constraints. One critical distinction is between metered and unmetered PDUs. Metered PDUs offer real time insights into power usage, both at the unit and outlet levels. This capability allows engineers to monitor power draw, identify inefficiencies, and plan for capacity growth. For example, a server consuming significantly more power than its peers may indicate a performance issue or the need for software optimization. The insights provided by metered PDUs help isolate such anomalies early and support proactive remediation.

In contrast, unmetered PDUs are simpler and more cost-effective but lack monitoring capabilities. While suitable for less critical or budget-constrained environments, they offer no insight into energy consumption or load distribution. In mission critical applications, the cost of not having insight into power usage can far outweigh savings from forgoing metering.

Thus, while unmetered PDUs may still serve a purpose, metered or intelligent PDUs are generally preferred in enterprise environments where uptime, energy management, and proactive maintenance are priorities.

PDUs also differ in terms of physical form factor and feature sets. Basic PDUs offer simple power distribution, while advanced PDUs include features such as remote outlet switching, power sequencing, and environmental monitoring (e.g., temperature and humidity sensors). The ability to remotely power cycle a server is especially valuable in large data centers, where physical access to racks may be delayed or restricted. This capability reduces mean time to repair (MTTR) and avoids unnecessary dispatches, streamlining incident response and minimizing service disruption.

The mounting orientation of PDUs either horizontal or vertical is another important consideration. Horizontal PDUs occupy rack unit (U) space within the equipment enclosure, whereas vertical PDUs (also known as zero U PDUs) are typically mounted along the rear side rails of the rack. Vertical configurations are space efficient and reduce cable congestion, promoting better airflow, simplifying cable management, and system maintenance. Improper PDU placement can disrupt cooling efficiency and lead to thermal hotspots, making optimal integration essential for both power and thermal performance.

Integrating PDUs with UPS systems and generator outputs requires a structured, safety-conscious approach. Lockout/Tagout (LOTO) procedures must always be followed before making any connections or adjustments, preventing accidental energization and reducing the risk of electrical hazards. Connection points are typically standardized, but engineers must still adhere to manufacturer guidelines and ensure proper grounding to maintain system integrity and personnel safety. Miswiring or failure to ground properly can result in damaging voltage imbalances, equipment failure, or even fire hazards.

Once operational, ongoing PDU monitoring is vital. Metered PDUs and integrated DCIM (Data Center Infrastructure Management) software can report power usage trends, identify abnormal consumption, and even flag devices that may be approaching overload conditions. For example, a slow and steady increase in current draw on a particular outlet could indicate a failing power supply or software inefficiency. The earlier such trends are identified, the more effectively engineers can respond before the issue results in downtime.

Power optimization within racks is a continuous process. Engineers can use PDU provided insights to implement power saving strategies, such as server consolidation, virtualization, or reconfiguration of workloads based on energy usage profiles. Grouping similar equipment with aligned power and cooling requirements allows for better airflow management, especially when high-power servers are placed near cold air inlets. These adjustments reduce cooling demands and lower overall operational expenditure.

Beyond individual server optimization, PDUs play a pivotal role in rack level and row level energy efficiency. Strategically managing which PDUs serve which racks helps distribute power evenly, avoid phase imbalances, and streamline maintenance workflows.

Advanced PDUs with remote switching can isolate faulty equipment or perform controlled power downs during maintenance windows without manual intervention. As data centers evolve toward automated, lights out operations, intelligent PDUs will become even more essential, serving as both control points and data sources for advanced infrastructure management.

In conclusion, PDUs are no longer passive distribution components they are intelligent, dynamic assets that underpin a resilient and efficient power delivery ecosystem. By enabling granular control, enhanced monitoring, and remote management, they empower engineers to make informed decisions that improve reliability, energy efficiency, and uptime. With careful selection, proper integration, and consistent oversight, PDUs make a significant contribution to the operational excellence of a data center.

Comprehensive Power Management and Preventative Maintenance in Data Centers

Effective power management in a data center hinges on a holistic understanding of the entire electrical infrastructure, encompassing generators, Uninterruptible Power Supply (UPS) systems, and Power Distribution Units (PDUs). These systems work in concert to provide uninterrupted, conditioned power to mission critical IT equipment. For a critical facilities engineer, mastery over the architecture, functionality, and maintenance requirements of each system is essential to minimizing downtime and maximizing operational efficiency.

The ability to monitor, manage, and troubleshoot these systems is a core competency. A facilities engineer must know how to interpret system data, detect anomalies, respond to alarms, and carry out interventions in real time. But beyond reactive efforts, proactive maintenance and systematic infrastructure planning are crucial for preventing failures. The implementation of robust preventative maintenance (PM) strategies ensures operational continuity, reduces emergency interventions, and contributes directly to cost control.

Preventative Maintenance: The Foundation of Reliability

Preventive maintenance is the foundation of a resilient and high performing data center power system. Unlike reactive repairs, PM involves scheduled inspections, systematic testing, cleaning, and proactive component replacement based on lifecycle data, not failure. Skipping or delaying maintenance can lead to catastrophic failures, data loss, and significant financial consequences.

The foundation of any PM program is a well defined maintenance schedule, tailored to the equipment's manufacturer guidelines, operational environment, and usage patterns. A layered approach daily, weekly, monthly, quarterly, and annual checks ensures thorough coverage:

- **Daily:** Visual inspection of UPS units and PDUs for signs of physical damage, unusual noises, or heat buildup.
- **Weekly:** Verification of battery voltage, UPS status indicators, and environmental sensor readings.
- **Monthly:** Inspection of electrical connections, ventilation paths, and component housing for dust or corrosion.
- **Quarterly/Semi-Annual:** Load testing, calibration of metering equipment, and firmware updates if required.
- **Annually:** Comprehensive battery testing, full system diagnostics, thermal imaging, and review of performance logs.

Battery Maintenance: Ensuring Backup Readiness

Batteries are essential to any UPS system, and their failure is one of the leading causes of UPS related outages. Regular testing and monitoring are crucial for evaluating battery health. For lead-acid batteries, this includes:

- **Voltage monitoring:** Ensuring each cell maintains charge within specified limits.
- **Specific gravity checks** (for flooded cells): Identifying electrolyte imbalances.
- **Load testing:** Simulating a power outage to verify actual performance.
- **Visual inspection:** Checking for swelling, corrosion, or leaking terminals.

As batteries age, their internal resistance increases, degrading their capacity. Proactive battery replacement, based on test results rather than failure, is a cost-effective strategy that avoids emergency downtime. In most cases, annual load testing is recommended, with increased frequency as the battery approaches the end of its service life. For lithium-ion systems, built-in Battery Management Systems (BMS) often provide predictive insights.

Electrical Connections and Safety Compliance

Every electrical joint in a UPS or PDU system is a potential point of failure if left unchecked. Loose or corroded connections can result in arcing, excessive heat, or fire hazards. During PM, engineers should:

- Use torque wrenches to confirm all connections meet manufacturer specifications.

- Inspect for oxidation or discoloration, especially in humid environments.
- Apply corrosion inhibitors when necessary.
- Confirm proper grounding, including bonding and earth connections.

Maintaining clean and secure electrical paths is essential for load stability and personnel safety. PM logs should include verification of torque values and photos of critical terminations.

Cleaning and Environmental Maintenance

While often overlooked, cleanliness directly impacts performance. Accumulated dust can clog vents, reduce airflow, and increase internal temperatures, potentially causing thermal shutdowns or a degraded lifespan. A robust PM program includes:

- Cleaning of fans, filters, and intake/exhaust vents using low pressure compressed air.
- Removal of any obstructions to airflow inside and around racks.
- Inspection and cleaning of heat sinks and internal enclosures following ESD safety protocols.

Environmental monitoring tools must be verified regularly. Sensors for temperature, humidity, and airflow should be tested and recalibrated if needed. The UPS should be located strategically to avoid proximity to HVAC exhausts or high heat areas.

Maintaining stable conditions typically 20–25°C with 40–60% humidity is essential for ensuring long term system performance and avoiding corrosion or battery degradation.

Logging, Analysis, and Continuous Improvement

An often underestimated part of preventative maintenance is documentation. Maintenance logs should include:

- Time stamped records of all activities performed.
- Test results (e.g., battery voltages, torque values).
- Observations and photos of anomalies or wear.
- Replacement dates and component serial numbers.

These records not only ensure compliance but also serve as the basis for predictive analytics. Over time, patterns in temperature fluctuation, power draw, or component degradation can inform adjustments to PM intervals and budgeting for replacements or upgrades.

By coupling this detailed, proactive maintenance approach with intelligent monitoring and real time alerting systems, critical facilities engineers can create a **self-reinforcing feedback loop** that continuously improves uptime, performance, and cost efficiency.

Regular testing of the UPS system's functionality is crucial. This involves performing simulated power outages to verify that the system automatically switches to battery power and that the transfer time is within acceptable limits. These tests should be performed according to the manufacturer's recommendations and with appropriate safety precautions. Testing should include verifying the UPS system's ability to provide power to the critical loads for an appropriate amount of time. This ensures that the system operates as intended and that critical IT equipment receives sufficient time to shut down gracefully during a power failure. Documenting the results of these tests provides valuable data that can be used to track the performance of the UPS system over time and to identify potential problems early on.

Preventive maintenance of the power distribution system extends beyond the UPS itself. This includes regular inspections and testing of PDUs, circuit breakers, and other related equipment. For example, visually inspecting the PDU for any loose connections, signs of overheating or damage, verifying that all power cords are securely connected, and checking the circuit breakers to ensure that they are properly functioning and appropriately sized for the loads they serve. Properly maintained power distribution equipment ensures that the UPS output can be successfully delivered to the IT equipment. An effective testing schedule includes performing routine maintenance checks on all the power distribution components. Failure to test these components can result in unexpected downtime and disruption to services.

Establishing and maintaining a comprehensive preventative maintenance program requires meticulous record keeping. All maintenance activities, including inspections, tests, and repairs, should be meticulously documented, along with the dates, times, and personnel involved. This documentation aids in identifying trends, predicting potential problems, and ensuring compliance with regulatory requirements. The data collected helps in optimizing the maintenance schedule and improving the overall efficiency of the maintenance program. Detailed records also assist in troubleshooting future problems, providing a historical context for understanding system performance and identifying patterns in equipment failures. These records can prove invaluable in situations such as insurance claims or audits.

Additionally, effective preventative maintenance depends greatly on the expertise and training of the personnel involved. Regular training programs should be put in place to ensure staff members are familiar with all procedures and safety protocols.

Hands on training, including practical demonstrations and simulations, is essential for developing the necessary skills and confidence to perform tasks safely and effectively. Ongoing training is crucial to stay current with evolving technologies and best practices. This training should cover all aspects of preventative maintenance, such as proper lockout/tagout procedures, safe handling of electrical equipment, and the use of specialized testing tools. Investing in well trained personnel is an investment in the overall reliability and uptime of the data center's power infrastructure.

Preventive Maintenance and Continuous Improvement Overview

Topic	Description
Maintenance Documentation	Logs should include time-stamped activities, test results (e.g., battery voltages), observations, photos, replacement dates, and serial numbers. Enables compliance, predictive analytics, and planning.
UPS Functional Testing	Simulated power outage tests ensure automatic battery switchover, acceptable transfer times, and sustained power to critical loads. Results should be documented for performance tracking.
PDU and Circuit Breaker Testing	Regular inspections of PDUs, circuit breakers, and power cords for loose connections, damage, and proper sizing. Ensures clean power delivery to IT equipment.

Record-Keeping and Trend Analysis	Detailed logs of inspections, repairs, and tests, including personnel names and timestamps, help identify trends, predict failures, and aid in troubleshooting or audits.
Training and Personnel Expertise	Staff should undergo ongoing training in LOTO procedures, equipment handling, safety protocols, and the use of testing tools. Hands-on and simulated training are key.
Strategic Investment Value	Proactive maintenance reduces downtime, extends equipment lifespan, and improves cost-efficiency. It's a long-term investment in infrastructure resilience.

By implementing a robust preventative maintenance program, encompassing regular inspections, testing, and cleaning, accompanied by meticulous record keeping and ongoing personnel training, data center operators can significantly improve the reliability and longevity of their power systems. This proactive approach minimizes costly downtime, maximizes equipment lifespan, and ensures the uninterrupted operation of critical IT infrastructure. The financial advantages far exceed the initial investment of time and resources, as proactive maintenance greatly lowers the risk of catastrophic failures and their associated costs.

Proactive maintenance is not simply a cost; it is a strategic investment in the long term success and resilience of the data center operation. A well structured preventative maintenance program is a cornerstone of effective data center management, contributing to a stable, reliable, and cost-effective operation.

Chapter 4: Fire Suppression and Security Systems

Fire Suppression in Data Centers: Strategy, Systems, and Safety

Fire suppression within a data center is a paramount component of critical facilities engineering, aimed at not only extinguishing potential fires but also minimizing damage to sensitive information technology (IT) infrastructure and ensuring uninterrupted business operations. The objective is to protect critical equipment, safeguard personnel, and maintain operational uptime, all while complying with stringent regulatory and safety standards. Achieving this necessitates a comprehensive understanding of available suppression technologies, their interactions within the data center environment, and the optimal integration within a broader fire protection strategy.

The selection of an appropriate fire suppression system is determined by a combination of factors, including the data center's size and configuration, equipment density, risk profile, and environmental considerations. Often, this process involves collaboration between critical facilities engineers, fire safety consultants, and data center architects to ensure that the chosen system balances effective fire mitigation with minimal impact on electronic equipment.

Gaseous Fire Suppression: Clean and Equipment-Friendly

Among the most preferred options in data center

environments are gaseous fire suppression systems. These systems suppress fire by displacing oxygen or inhibiting the chemical reactions involved in combustion, without leaving behind water or residue that might damage electronics. Common gaseous agents include:

- **Argonite:** A fast acting inert gas, effective but requiring large storage volumes.
- **Inergen:** A blend of nitrogen, argon, and CO_2 that suppresses fires while maintaining enough oxygen for human safety.
- **FE-13:** A halocarbon agent known for quick discharge and low environmental impact.

These gases are non conductive and non corrosive, making them ideal for high value IT environments. The systems are deployed through precision-engineered nozzles installed throughout the room, ensuring even dispersion for maximum suppression coverage.

However, gaseous systems introduce a critical occupant safety consideration as oxygen levels drop, rapid evacuation is necessary. Proper signage, audible and visual alarms, and emergency egress plans must be in place. In high occupancy areas, Inergen may be preferred over Argonite due to its safer oxygen retention characteristics.

Regular maintenance and testing of gaseous systems are essential. Key tasks include:

- Verifying cylinder pressure and nozzle integrity.
- Conducting discharge simulation tests.
- Inspecting piping and control panel connections.

- Logging all maintenance and test outcomes for compliance auditing.

Water-Based Suppression

Though traditionally avoided due to the risk of water damage, water based fire suppression systems still play a role in certain data center zones, especially where fire risk is elevated. Systems such as pre action sprinklers and deluge systems have been adapted for data center use, featuring enhanced control features.

- Pre action systems require two triggers (such as heat and smoke detection) before water is released. This provides an additional buffer, reducing the risk of accidental water discharge.
- Deluge systems, in contrast, activate rapidly and are used in high-hazard areas where quick fire spread is a concern.

To mitigate water damage risks, water based systems should be designed with:

- **Fine-spray, low-pressure nozzles** to limit water volume.
- **Strategically positioned sprinklers**, avoiding direct discharge over critical hardware.
- **High-capacity drainage systems** are used to quickly remove discharged water from the area.

These features help strike a balance between fire suppression efficacy and equipment protection.

Fire Detection: The First Line of Defense

No suppression system can function effectively without

early detection. Integrated fire detection systems act as the first alert, triggering suppression mechanisms and enabling personnel to evacuate safely. Detection systems commonly consist of:

- **Smoke detectors** (both ionization and photoelectric types) for fast detection of smoldering or flaming fires.
- **Heat detectors** to capture rapid temperature changes in areas with low airflow.
- **Flame detectors** (using infrared or ultraviolet sensors) to detect open flames.

Detector placement is strategic and layered, with a higher density in zones like server racks, electrical panels, and cable runs, which pose increased fire risks. It's critical to avoid coverage gaps (dead zones) and ensure airflow patterns don't interfere with detector performance.

Regular testing and maintenance of these systems include:

- Manual and automated functionality checks.
- Calibration of sensor sensitivity.
- Verification of alarm communication with building fire control systems.
- Clear documentation of test results and repairs.

Alarms must interface directly with the building's fire alarm control panel and be tied into automated emergency response protocols. This ensures immediate notification to fire personnel, triggers suppression, and initiates facility wide alerts.

Integrated Response Strategy

A comprehensive fire protection strategy in a data center isn't just about extinguishing fires it's about prevention, containment, and recovery. The engineer's role encompasses:

- Performing risk assessments to identify vulnerable zones.
- Coordinating suppression system design with HVAC, electrical, and layout planning.
- Developing evacuation and safety protocols specific to fire suppression systems.
- Training personnel in system operations and emergency response.

Ultimately, documentation is critical. A well maintained fire protection log should include:

- System test records.
- Detector maintenance history.
- Suppression system inspections.
- Evacuation drill outcomes and revisions.

Integrated Fire Suppression and Detection in Data Centers

Fire suppression in data centers is not simply about extinguishing flames it is a critical component of infrastructure protection, business continuity, and life safety. Effective fire protection systems must be meticulously designed to minimize disruption to sensitive IT equipment, relying on an intricately coordinated network of detection, suppression, control, and response mechanisms. These systems must adhere to stringent regulatory standards while maintaining exceptional responsiveness and reliability in real world scenarios.

A comprehensive fire protection strategy seamlessly integrates both suppression and detection systems. The fire detection system serves as the initiator of the suppression process and often activates additional emergency responses, such as ventilation shutdowns, power isolation, and building evacuation protocols. This integrated functionality necessitates flawless coordination among sensors, control panels, alarm systems, and communication networks. Typically, a centralized monitoring station oversees this system, receiving input from various detectors and executing appropriate responses based on pre established thresholds and emergency protocols.

Proper integration requires rigorous testing and validation to confirm that systems interact as intended. Simulated fire scenarios are essential for validating response timing, coverage, and system communication. Personnel training is equally vital operators must understand how the system functions, including both its capabilities and limitations. Training should combine theoretical instruction with practical drills, helping personnel build confidence in emergency procedures and system diagnostics.

Gaseous and Water-Based Suppression Systems

Gaseous suppression systems are a preferred solution in IT environments due to their non conductive, residue free nature. They suppress fire by displacing oxygen or interrupting the combustion process, all while leaving no water damage behind. Common agents like Argonite, Inergen, and FE 13 each offer unique properties. Argonite and Inergen, being inert gases, lower the oxygen concentration in the room to a level that extinguishes fire but is marginally safe for humans. FE 13, a halocarbon agent, suppresses fire quickly and is well regarded for its low environmental impact.

System design for gaseous suppression involves strategic nozzle placement and precise calculations of gas quantity based on room volume and required concentration levels. Safety mechanisms ensure delayed discharge, allowing for evacuation before oxygen displacement becomes hazardous. These systems must be tested regularly to confirm cylinder pressure, valve integrity, and nozzle alignment. Any deficiencies discovered during testing must be documented and addressed immediately to ensure full readiness.

Water based systems, although riskier in IT environments, are still used in some applications. Pre action sprinkler systems which require a confirmed fire signal before releasing water help lower the risk of false discharges. Deluge , which respond more aggressively, release water quickly and are usually installed in high risk areas. To minimize damage, these systems can be fitted with fine spray, low pressure nozzles and supported by dedicated drainage systems to direct water away from sensitive equipment.

Fire Detection Technologies and Placement

Detection is the first line of defense in fire mitigation. Modern systems use a combination of smoke, heat, and flame detectors to ensure layered, reliable detection across diverse operating conditions.

- **Smoke detectors** are foundational to early fire warning. Ionization detectors are more sensitive to fast-flaming fires that emit small smoke particles, while photoelectric detectors excel in detecting slow, smoldering fires that produce larger smoke particles. Choosing the appropriate detector depends on the specific risks and airflow conditions of each room.

- **Heat detectors** respond to elevated temperatures. Fixed temperature types activate at a specific threshold, while rate rise detectors trigger based on rapid temperature increases. These are especially useful in dusty or high airflow areas where smoke detectors may underperform.
- **Flame detectors**, using infrared or ultraviolet sensors, offer fast detection of visible flames and are typically used to supplement other detector types in areas where open flame risk is high.

Detector placement is a highly strategic process. Areas with high risk, such as server rooms, power rooms, and cable vaults, need dense coverage. However, deploying too many detectors can cause false alarms and add unnecessary system complexity. Fire modeling software aids in optimizing detector placement by considering room shape, air flow, and obstructions to ensure maximum coverage and reduce dead zones.

Alarm Systems, Monitoring, and Engineering Response

An effective alarm system not only alerts on site personnel but also interfaces with building wide emergency systems and external agencies. Alarm systems must offer clear indicators of fire location, type, and severity—ideally in real time via both visual displays and audible alarms. Many systems now support remote monitoring, allowing centralized supervision across multiple sites and enhancing situational awareness through integrated dashboards.

In emergency situations, the critical facilities engineer plays a key role. Upon receiving an alarm, they must assess the nature of the threat, verify alarm accuracy, initiate evacuation protocols, and ensure the suppression system activates properly. They also liaise with fire departments and security teams while monitoring system performance during the event. Following the incident, the engineer assists in investigating the root causes, evaluating system performance, and recommending improvements to hardware, layout, or protocols.

Maintenance, Compliance, and Continuous Improvement

Preventative maintenance ensures that detection and suppression systems remain reliable over time. This includes:

- **Functional testing** of smoke, heat, and flame detectors to ensure response time and sensitivity remain within tolerance.
- **System-wide testing** to validate the response sequence from detection to suppression and notification.
- **Inspection of control panels, wiring, and sensor connections** for damage, corrosion, or tampering.
- **Cleaning and calibration** of detectors to eliminate false alarms caused by dust or obstructions.

All activities must be meticulously documented, creating an auditable history of testing, maintenance, and corrective actions. Records should include test dates, results, the personnel involved, and any subsequent follow up measures.

Regulatory compliance is non-negotiable. Data centers must adhere to local and international fire codes (e.g., NFPA 75, NFPA 2001), which govern system design, maintenance frequency, and inspection protocols. Periodic audits by fire safety authorities help confirm compliance and expose areas for improvement. Utilizing technologies such as remote monitoring, data analytics, and fault trend tracking further improves system performance and responsiveness.

Fire suppression and detection in data centers involve more than just having extinguishing agents and alarms they require an integrated design, precise implementation, rigorous maintenance, and skilled personnel. The critical facilities engineer plays a central role in this system, making sure all components work together smoothly and responses are quick and effective. A well maintained, properly integrated fire protection system is not only a regulatory necessity but also a key element of business continuity, safeguarding lives, data, and equipment equally.

Fire Suppression and Detection in Data Centers

Effective fire suppression and detection are fundamental to critical facilities engineering, safeguarding not only sensitive IT equipment but also ensuring operational continuity. In a data center environment where even minor disruptions can result in significant data loss or financial damage, fire protection must be both fast and non-intrusive. The goal is not simply to extinguish a fire, but to do so with minimal impact to equipment, infrastructure, and personnel. Achieving this requires a layered strategy that integrates detection systems, suppression mechanisms, and procedural coordination, all tailored to the specific risks and configuration of the data center.

One of the most widely implemented fire suppression strategies in modern data centers is the use of gaseous fire suppression systems. These systems employ inert gases such as Argonite, Inergen, or FE 13 to reduce the oxygen concentration in a room, effectively suffocating a fire while leaving no water or corrosive residue behind. This makes them ideal for environments where water based systems could cause irreversible damage to electronics. Each gas agent has specific properties that influence system design. For instance, Argonite, while effective, requires a larger storage volume. Inergen, a blend of nitrogen, argon, and CO_2, is environmentally friendly and human safe at lower concentrations. FE 13 offers rapid suppression with relatively low toxicity and environmental impact.

The architecture of gaseous systems typically includes a central bank of gas cylinders connected to a network of pipes and nozzles distributed throughout the protected area. Upon activation, the agent is released and evenly dispersed to achieve the required concentration for fire suppression. These systems must be carefully engineered based on room size, ceiling height, ventilation rates, and the types of materials present. Because these gases displace oxygen, human safety protocols are essential. Evacuation procedures must be in place, and proper signage, alarms, and training are mandatory to ensure personnel can exit safely before discharge.

Routine maintenance and inspection are vital for gaseous systems. This includes checking gas cylinder pressure, inspecting nozzle alignment, and verifying system integrity through diagnostic checks. Cylinders must be replaced or recharged after discharge or when pressure levels fall below thresholds. Preventive testing ensures that no leaks or blockages exist in the pipework and that all actuators and release mechanisms are fully operational.

Though less common in high density environments, water based suppression systems still play a role in specific parts of a data center or legacy facilities. Modern adaptations, such as pre action sprinkler systems, provide a middle ground. These systems require both a fire detection signal and a manual or secondary verification before water is released, reducing the risk of accidental discharge. Another variation, deluge systems, are designed for rapid deployment in high risk areas like fuel storage or generator rooms, where intense fires could spread quickly. Both pre action and deluge systems rely heavily on detection input and zoning logic to prevent collateral damage to IT equipment. Specialized low flow or mist nozzles can also minimize water use and lower the risk to electronics.

For water based systems, drainage design is critical. Any discharge must be channeled away from critical equipment through well designed flooring and runoff systems. Regular maintenance includes pressure testing, valve checks, and flushing to prevent sediment buildup or corrosion within the pipes. Sprinkler heads must be inspected for blockages or alignment issues, and water sources must be reliable and adequately pressurized.

Whether gaseous or water based, fire detection systems form the backbone of the suppression strategy. These systems provide early warning of fire conditions, enabling swift response and triggering the appropriate suppression method. A combination of smoke, heat, and flame detectors offers layered detection capabilities tailored to different types of fire risk.

Smoke detectors are typically the first line of defense and come in several forms. Ionization detectors are more responsive to fast flaming fires, as they detect the disruption of ionized air currents by fine smoke particles. Photoelectric detectors, on the other hand, are more effective at identifying smoldering fires, using a light beam to sense the scattering caused by larger smoke particles. The choice between these technologies depends on the layout and specific fire risks present in each room. In many cases, both types are installed in tandem to maximize coverage and response capability.

Heat detectors provide valuable redundancy, particularly in areas with high airflow or particulate matter where smoke detectors may be less effective. These include fixed-temperature detectors, which activate when ambient temperatures exceed a certain threshold, and rate rise detectors, which respond to rapid increases in temperature. Their presence helps confirm fire presence in less accessible or dust prone areas.

Flame detectors, which use infrared (IR) or ultraviolet (UV) sensing, are highly effective for detecting open flames. These devices are especially useful in areas where fast, visible flames may arise, such as fuel storage or generator rooms. However, due to their susceptibility to false positives in certain environments, they are usually paired with other detection technologies to provide confirmation before activating suppression systems.

The placement and zoning of all detectors is a science in itself. Detectors must be strategically distributed based on factors such as room airflow, ceiling height, equipment density, and potential fire hazards. High risk zones like server racks, power distribution areas, or battery rooms require increased detector density. Fire modeling software is often used during system design to simulate smoke dispersion and optimize detector placement, ensuring coverage without excessive overlap or nuisance alarms.

A comprehensive alarm and communication system integrates all detection elements and communicates alerts to facility personnel as well as external emergency responders. Contemporary detection systems are often integrated into centralized Building Management Systems (BMS) or Data Center Infrastructure Management (DCIM) platforms. These capabilities facilitate remote monitoring, real-time data visualization, and event logging, ensuring that any incident is swiftly recognized and addressed. Upon activation of an alarm, these systems may also automatically initiate additional protective measures, such as deactivating HVAC systems, closing fire dampers, or isolating power supplies to affected areas.

Central to an effective response is the integration between detection, suppression, and response systems. The detection system must interface with suppression controls, emergency lighting, evacuation systems, and even ventilation and power shutdown protocols. These systems must be synchronized through a well designed control panel and verified through commissioning tests and periodic simulations. Many facilities use networked addressable detection systems that can localize the source of a fire to a specific room or rack, improving response efficiency.

Fire Suppression and Detection Overview

Topic	Description
Purpose of Fire Systems	Protect sensitive IT equipment and maintain operational continuity with fast, low-impact fire response.
Gaseous Suppression Systems	Use inert gases (Argonite, Inergen, FE-13) to reduce oxygen and suppress fire without residue. Safe for electronics, it requires careful room-specific design and adherence to human safety protocols.
Gaseous System Maintenance	Includes pressure checks, nozzle alignment, leak detection, and actuator testing. Cylinders must be recharged or replaced after they have been discharged.
Water-Based Systems	Used in some areas or legacy setups. Includes pre-action (double-confirmation before discharge) and deluge systems (rapid deployment). Water mist or low-flow nozzles minimize IT risk.
Water System Maintenance	Includes drainage planning, pipe flushing, valve and pressure checks, and sprinkler head inspections.
Detection Technologies	Layered detection: smoke (ionization & photoelectric), heat (fixed temp & rate-of-

	rise), and flame detectors (UV/IR). Chosen based on room conditions and fire risks.
Detector Placement and Zoning	Detectors are placed based on airflow, ceiling height, equipment density, and risk zones. Fire modeling helps optimize coverage.
Alarm & Communication Systems	Tied into BMS/DCIM platforms for real-time alerts, logging, and action initiation (e.g., shut down HVAC, isolate power).
System Integration	Detection, suppression, alarms, evacuation, and shutdown systems must interface and be tested together. Addressable systems enhance response precision.

For these integrations to work in real time, trained personnel are critical. All critical facilities engineers and fire safety teams must be proficient in the system's operation, including manual overrides, emergency procedures, and reset protocols. Training should include both theoretical knowledge and practical drills. Simulated fire events can help familiarize staff with system behavior under real conditions, improving decision-making and reducing panic during actual incidents.

Ensuring regulatory compliance is a critical component of fire protection in data centers. Compliance with local fire codes and international standards, such as NFPA 75, NFPA 76, NFPA 2001, as well as guidelines from ISO and ASHRAE, informs system design, performance metrics, and inspection protocols. Regular audits conducted by both internal safety teams and external certifying authorities ensure that fire suppression systems are fully operational and legally compliant. It is imperative that every testing, inspection, and maintenance action is meticulously documented, creating a comprehensive paper trail for audits, insurance validation, and post incident analyses.

Routine system testing should encompass verification of the functionality of individual detectors, pressure checks on both gas and water systems, confirmation of alarm signals, and execution of full system simulations. This includes quarterly and annual functional testing as prescribed by the manufacturer and regulatory guidelines. Detectors should also be cleaned periodically to prevent false alarms or sensor drift, particularly in high dust environments.

Ultimately, a proactive and integrated fire protection program is more than just a safety obligation; it is a fundamental pillar of data center uptime and business continuity. While the incidence of fire risks is statistically low in well-managed facilities, the potential for catastrophic losses underscores the necessity for comprehensive systems and processes to address these risks effectively.

By combining advanced detection technologies, appropriate suppression strategies, regular training, and strict regulatory adherence, critical facilities engineers can ensure their infrastructure remains resilient, responsive, and secure against one of the most serious threats in any IT environment.

Physical Security and Emergency Response Planning

Securing a data center's physical infrastructure requires more than basic deterrents it demands a robust, multi-layered strategy that integrates access control, surveillance, intrusion detection, and well practiced emergency response protocols. The critical facilities engineer plays a central role in this ecosystem, tasked with maintaining the integrity, responsiveness, and ongoing evolution of all security systems.

Access control is the primary line of defense, ensuring that only authorized personnel can access secure areas within the data center. Conventional key card systems are commonly used, allowing access based on role specific clearance levels. Proximity readers and integrated time tracking improve efficiency and accountability. Advanced systems may utilize biometric authentication such as fingerprint scanners or iris recognition—to minimize the risk of credential theft and to keep detailed logs of each access attempt. However, maintaining system integrity requires more than just installation; regular audits are crucial. As personnel roles evolve or personnel leave, access rights must be updated promptly, outdated credentials revoked, and logs reviewed for any suspicious or unauthorized activity.

Closely tied to access control are surveillance systems, particularly closed-circuit television (CCTV). Strategically placed high resolution cameras, capable of pan tilt zoom functionality, monitor both internal and external points of entry. Their data is managed through centralized video management systems (VMS), which allow real-time and recorded footage review. Surveillance layout must be carefully planned to eliminate blind spots and ensure full coverage of sensitive zones such as server rooms, electrical closets, and loading docks. Adequate lighting is essential to maximize visibility and footage clarity. The engineer must also ensure that video data storage meets retention requirements for audits or forensic analysis in the event of a breach.

To augment access and surveillance, intrusion detection systems (IDS) monitor for unauthorized activity using motion sensors, door contacts, vibration detectors, and pressure mats. Upon detecting irregular activity, these sensors communicate with the central monitoring system, which can trigger alarms and initiate a pre defined response protocol. Integrating IDS with access logs and CCTV provides a multi dimensional picture of events, enabling swift verification and response. Calibration is important environmental factors like airflow or wildlife can trigger false alarms, so sensors must be appropriately tuned.

Beyond routine maintenance, the critical facilities engineer is also the first responder to security alarms and incidents. When an alert is received, the engineer must verify its validity, assess potential threats, and initiate the proper protocol. This may involve coordinating with on site security, analyzing surveillance footage, contacting law enforcement, or activating lockdown procedures. After any incident, the engineer conducts a detailed investigation, identifies vulnerabilities, and suggests improvements to the security system. Proper documentation of incidents is crucial, including event timelines, response actions, damage assessments, and system performance, to help prevent future issues.

Part of this ongoing security oversight involves collaboration with external security providers and technology vendors. Engineers often coordinate security audits, oversee upgrades, and help integrate emerging technologies to enhance the data center's security posture. These relationships ensure the facility benefits from up to date expertise while maintaining regulatory compliance with industry standards such as ISO/IEC 27001 or SSAE 18. The critical facilities engineer thus bridges internal operations with external best practices, forming a vital link in the center's overall security resilience.

While securing the data center is essential, so too is preparing for the unexpected. Emergency response planning is not a one time task, but a dynamic, evolving strategy designed to mitigate the impact of incidents ranging from minor equipment failures to large scale disasters. A resilient plan is characterized by clear procedures, well defined roles, and continuous training.

Effective planning begins with a comprehensive risk assessment. This means not just identifying potential hazards such as electrical fires, prolonged power loss, HVAC failures, or physical intrusions but also evaluating the likelihood and potential consequences of each. Data centers in flood prone areas, near seismic fault lines, or in urban zones with elevated crime rates face different sets of risks. The assessment should also account for operational dependencies and business critical systems. Only with this granular understanding can appropriate priorities, resources, and redundancies be built into the response plan.

From this risk matrix, the engineer develops detailed response procedures tailored to each threat. These should clearly outline every action to be taken, from the moment a threat is detected to the return to normal operations. For instance, a fire response might involve initiating the alarm, evacuating personnel, triggering fire suppression systems, and notifying emergency services. Specific roles must be assigned to personnel, including those who communicate with first responders, those who secure the infrastructure, and those who monitor system performance during the incident. Visual aids such as color coded flowcharts or laminated quick reference guides can improve clarity during high stress moments.

Central to any emergency response is communication. Internal protocols must establish clear chains of command and define communication channels, whether via two-way radios, SMS alert systems, pagers, or dedicated hotlines. Externally, the data center must coordinate with municipal emergency services, facility management, and stakeholders, including customers or tenants. The engineer is responsible for ensuring these systems are tested regularly and documented clearly, with redundant pathways in place to ensure functionality even during infrastructure failures.

Testing is as crucial as planning. Regular emergency drills and simulations ensure that everyone understands their role and that systems work as expected. These drills should differ in scope and complexity, from partial walk throughs to full-scale simulations that imitate real scenarios like a fire outbreak, generator failure, or data breach. Each exercise should end with a structured debrief, during which gaps, delays, or miscommunications are identified. Outcomes are then used to improve procedures and train new staff. Over time, this iterative process enhances institutional readiness and integrates emergency response protocols into daily operations.

Certain emergencies, such as power outages, require special attention due to their potential to trigger cascading failures. A dedicated procedure must outline the steps from detection to resolution, including verifying utility status, initiating backup power through the UPS and generator systems, notifying stakeholders, and prioritizing the safe shutdown of non essential systems. Care must be taken to ensure data integrity throughout this process, including staged server shutdowns and pre established power down sequences. The restoration process should include systematic inspections, system checks, and environmental monitoring before returning to normal operations. These procedures must be tested and refined just like any other, with simulations designed to evaluate the facility's full blackout recovery capabilities.

Together, physical security and emergency response form the twin pillars of operational resilience. The critical facilities engineer's responsibilities in this domain extend far beyond passive oversight they are actively engaged in system design, threat assessment, incident response, and continuous improvement. Through the integration of layered technologies and human procedures, the engineer helps create a secure, agile, and responsive data center environment. This vigilance ensures the safety of both personnel and infrastructure, protects against costly downtime, and upholds the trust that clients place in the reliability of the facility.

Equipment failures are another potential disruption that requires a detailed response. The plan should include specific procedures for each type of equipment, outlining steps to isolate the faulty component, perform diagnostics, initiate repairs, and implement temporary workarounds if needed. A detailed inventory of spare parts and replacement equipment is also essential to minimize downtime. This includes defining maintenance procedures and the process for obtaining replacement components. The plan must also address communication protocols for notifying relevant parties, including tenants and vendors. It should include provisions for handling incidents that may involve external emergency services, such as fire departments, police, or medical teams. Establishing clear communication procedures and designating a point of contact for external communication are crucial. The plan should also detail the data center's location, access points, and any potential hazards, ensuring this information is easily accessible to emergency responders. Creating detailed maps and diagrams showing the location of critical equipment, fire suppression systems, and emergency exits can be extremely helpful for responders. In conclusion, an effective emergency response plan is more than a document; it's a dynamic, living framework that requires ongoing review, refinement, and practice. By combining thorough risk assessment, meticulous planning, regular drills, and clear communication protocols, data center operators can greatly improve their ability to respond effectively to incidents of all sizes, minimizing disruption and protecting valuable assets. Investing in this preparedness is not just a cost but an investment in the ongoing operational success and safety of the business. Additionally, a strong emergency response plan demonstrates due diligence, fulfills regulatory requirements, and enhances the facility's resilience against unforeseen events. The goal is not only to reduce the impact of incidents but also to learn from each one to strengthen future preparedness.

Maintaining the integrity and functionality of a data center's security systems is paramount. A robust security posture isn't solely about preventing physical intrusions; it's about establishing a layered defense that anticipates and mitigates a broad spectrum of threats, from unauthorized access to sophisticated cyberattacks that could have physical ramifications. This section delves into the crucial aspects of monitoring and maintenance, emphasizing preventative measures to minimize the risk of security breaches. Neglecting these procedures can lead to substantial financial losses, reputational damage, and even legal repercussions.

Regular inspections form the backbone of proactive data center security management. These reviews must go beyond cursory visual checks and instead require systematic, methodical examinations of each component within the security infrastructure. Access control systems demand detailed attention engineers must verify the functionality of card readers, check the physical integrity of locks and door alignment, and audit access databases for inconsistencies. Anomalies, such as malfunctioning readers, forced entries, or irregular access patterns, must be investigated immediately. Each inspection should be thoroughly documented, with logs detailing the date, time, and any identified issues. These records not only support troubleshooting but also serve as critical evidence in security audits or investigations.

While physical hardware is important, the software and network infrastructure supporting these systems require equal scrutiny. Regular updates to access control software are vital to address vulnerabilities and maintain compatibility with modern security standards. These updates must be implemented in a controlled, minimally disruptive manner. Simultaneously, the associated network infrastructure should undergo periodic audits to detect unauthorized access points, verify firewall configurations, and ensure no latent security holes exist. Penetration testing—often performed by cybersecurity specialists—can simulate real world attacks and reveal critical flaws. Findings from these tests inform corrective actions and long term system hardening strategies.

CCTV systems also require consistent, detailed maintenance. These systems provide visual records of internal and external activity, forming a key layer of security. Routine inspections must confirm that each camera functions properly, produces clear images, and maintains a continuous connection to its recording device. Maintenance tasks include cleaning lenses, checking for tampering, and verifying that field view angles remain optimal. Camera placement should be reevaluated periodically to eliminate any developing blind spots as layouts or equipment configurations evolve. Engineers should also test the recording system's storage and backup capacity to ensure footage is retained according to regulatory requirements and is protected from loss or corruption.

Physical Security and Emergency Response Planning Overview

Topic	Description
Access Control Systems	Use key cards, proximity readers, and biometrics for role-based entry. Requires regular audits and timely revocation of outdated credentials.
Surveillance (CCTV)	High-resolution, strategically placed cameras monitored via VMS. Ensure full coverage, retention, and clarity; routinely inspect and test systems to maintain optimal performance.
Intrusion Detection	Motion sensors, door contacts, and vibration detectors trigger alerts, which

Systems (IDS)	are integrated with access logs and CCTV for verification. Must be tuned to avoid false alarms.
Engineer's Security Role	Acts as first responder for alerts, verifies incidents, coordinates with security/law enforcement, and conducts investigations and follow-ups.
Vendor Coordination & Compliance	Engineers liaise with external vendors, conduct audits, and ensure systems meet standards (e.g., ISO/IEC 27001, SSAE 18).
Risk Assessment	Identify and evaluate potential threats (e.g., fires, power failures, intrusions) by location and facility type. Tailor response plans accordingly.
Emergency Procedures	Define actions for each scenario (e.g., fire, intrusion), assign roles, use visual aids for clarity, and establish escalation protocols.
Emergency Communication Systems	Define internal and external communication paths (e.g., radios, SMS, hotlines). Ensure redundancy and regular system testing.
Emergency Drills & Simulations	Conduct full or partial drills to practice roles, verify system performance, and identify weaknesses. Use findings to

	update response plans.
Power Outage Protocols	Include utility verification, UPS/generator activation, stakeholder notification, and safe shutdown of non-essential systems. Test regularly.
Equipment Failure Response	Outline steps to isolate, diagnose, repair, or replace failed components. Maintain spare inventory and notify stakeholders as needed.
Coordination with Emergency Services	Establish contact protocols, share maps, hazard data, and access points with responders to ensure effective communication and coordination. Assign liaison personnel.
Security Monitoring and Maintenance	Inspect access control, locks, logs, and alarm systems. Investigate anomalies. Maintain accurate, time-stamped records.
Security Software and Network Health	Update software regularly, perform audits and penetration testing to identify vulnerabilities. Secure network infrastructure.
CCTV System	Verify image quality, connections, and camera angles to ensure optimal

Maintenance	performance. Clean lenses, test storage/backup, and eliminate blind spots.

Intrusion detection systems (IDS) provide yet another essential line of defense, using motion detectors, vibration sensors, and contact switches to detect unauthorized entry attempts. Integrated with CCTV and access control platforms, these systems offer real time alerts and incident correlation. Maintenance of IDS includes verifying the responsiveness and sensitivity of each sensor and confirming that the alarm logic functions as intended. False alarms are a common issue and must be addressed through careful calibration and environmental adjustment, reducing nuisance triggers without compromising security. Review of system logs helps to identify recurring anomalies or weaknesses, providing insight for system refinement and physical security adjustments.

Effective security extends beyond individual components and depends heavily on seamless system integration. Sophisticated analytics platforms now enable real time coordination among access control, CCTV, and IDS. These systems aggregate and correlate data to identify suspicious patterns, such as repeated access attempts, abnormal movements, or unauthorized entry after hours. Regular functional testing ensures that these integrated platforms operate without error and that their automated alerting functions work correctly. Backup procedures and failover capabilities must also be regularly tested to confirm that data will be preserved and accessible in the event of unexpected failures or cyberattacks.

Real world failures show the high cost of ignoring proactive security maintenance. In one notable case, a data center experienced a breach because it used outdated access control software that let attackers bypass authentication protocols. This resulted in unauthorized access to critical servers, causing a costly data breach and damage to its reputation. In another incident, a neglected CCTV system failed to record a theft because a camera malfunction went unnoticed during routine checks. These examples highlight the importance of consistent maintenance, as even small oversights can lead to serious consequences.

Keeping pace with technological evolution is also essential. Security systems that were state - art five years ago may now be outdated and vulnerable. Strategic upgrades should include advanced biometric access controls, AI enhanced video analytics, and intrusion detection technologies that respond to modern threats. These improvements should be implemented carefully, ideally using a phased deployment approach that allows for rigorous testing and minimal disruption to operations. Compatibility with existing infrastructure should be assessed early in the planning process to prevent integration challenges and preserve data integrity during transitions.

Equally important is the training and preparedness of personnel tasked with operating and monitoring these systems. Engineers and security staff must be proficient in each system's use, able to interpret alarms accurately, and capable of responding swiftly and effectively. Training should not only cover technical operations but also include simulations and hands on exercises that replicate real world security scenarios. Personnel must understand both procedural protocols and the underlying rationale behind system behavior, enabling them to make more informed decisions under pressure. Developing analytical skills is especially important for identifying subtle behavioral patterns and correlating disparate alerts into meaningful, actionable insights.

Ultimately, proactive security maintenance is a key investment in the long-term resilience and reliability of the data center. By thoroughly inspecting systems, keeping detailed logs, upgrading infrastructure, and training staff, operators can significantly lower the risk of breaches, prevent costly incidents, and preserve stakeholder trust. The expense of preventive maintenance is minimal compared to the financial and reputational damage caused by a security failure. In a landscape where threats are constantly changing, maintaining a layered, proactive, and adaptable security approach is not just best practice it is a business necessity.

The goal is not just to react to threats but to build a system that deters, detects, and neutralizes them before they escalate, ensuring the uninterrupted protection of both physical and digital assets.

Chapter 5: Data Center Efficiency and Optimization

Power Usage Effectiveness

Understanding Power Usage Effectiveness (PUE) is essential for optimizing data center efficiency. As a widely accepted metric, PUE quantifies how effectively a data center uses energy by comparing the total facility power consumption to the power consumed specifically by IT equipment. A lower PUE indicates greater energy efficiency and cost-effectiveness, aligning closely with sustainability goals. Although the formula is straightforward, its interpretation requires a nuanced understanding of infrastructure and operations.

The basic formula is:

PUE = Total Facility Power / IT Equipment Power

Total Facility Power includes all energy used in the data center, such as cooling systems (like CRAC units and chillers), power distribution (PDUs), lighting, security systems, and other support equipment. IT Equipment Power specifically refers to the power used by servers, storage devices, and network hardware. Accurate PUE calculations rely on precise measurements of both inputs, typically gathered through strategically placed power meters and complemented by real-time data from Data Center Infrastructure Management (DCIM) platforms.

Interpreting PUE requires context. A perfect PUE of 1.0 would mean that all energy is used solely by IT hardware, which is virtually unattainable due to the necessary overhead of cooling and lighting. In practice, a PUE below 1.2 is considered excellent; 1.2 to 1.5 is good; anything above 1.5 suggests there's significant room for efficiency improvement. However, benchmarks vary depending on climate, facility design, and hardware type. For example, data centers in hotter climates often have higher PUEs due to increased cooling demands.

Monitoring PUE over time is critical for identifying inefficiencies and guiding improvements. A rising PUE might signal deteriorating cooling performance, increased equipment aging, or poor power distribution. Conversely, a declining PUE reflects successful energy management. Analyzing PUE alongside other performance indicators enables operators to proactively diagnose issues and fine tune their operations, thereby reducing both costs and environmental impact.

Multiple factors affect a facility's PUE. Chief among them is cooling system efficiency. Outdated or oversized cooling units increase energy consumption, especially if airflow is poorly managed. Hot spots caused by misaligned racks or blocked air paths often result in overcooling. Modern solutions like liquid cooling or free air cooling provide significant benefits. Best practices such as hot aisle/cold aisle containment and optimized server rack layout further improve airflow and cooling efficiency.

Power distribution infrastructure also plays a critical role. Inefficient or aging PDUs contribute to power loss, raising overall consumption. Replacing them with modern units that offer power monitoring and load balancing improves energy distribution and reduces waste. Strategically spreading loads across PDUs ensures even energy usage, preventing circuit overloads and reducing unnecessary strain on power components.

The choice and condition of IT hardware are equally important. Older servers usually lack energy saving features and consume more power. Upgrading to newer, energy-efficient models—especially those designed with power management in mind can significantly lower the IT workload. Virtualization and server consolidation efforts help reduce unnecessary energy use, while decommissioning outdated equipment ensures energy isn't wasted on idle or underused systems.

Operational practices also have a direct impact on PUE. Smart capacity planning prevents the inefficiencies associated with overprovisioning. Routine maintenance of both IT and infrastructure equipment helps preserve energy efficiency, while robust monitoring systems detect anomalies before they escalate. Automated shutdown policies for idle hardware, along with enforcing power management settings, further support energy conservation.

Real world examples clearly demonstrate the importance of PUE. A legacy data center relying on outdated cooling and disorganized airflow might operate with a PUE of 2.0 or higher. In contrast, a well designed facility utilizing advanced cooling techniques, virtualized servers, and real time energy monitoring may maintain a PUE below 1.5. The operational cost and environmental impact differences between these facilities can be substantial, reinforcing the value of strategic upgrades.

Lowering PUE requires a multifaceted approach. Key strategies include replacing inefficient cooling systems and PDUs, adopting advanced cooling methods like liquid or free air cooling, and improving airflow through better rack placement and containment strategies. Equally important is deploying energy-efficient IT hardware and consolidating workloads via virtualization. Regularly retiring obsolete equipment ensures that only essential devices consume power. Together, these strategies promote a leaner, greener, and more resilient data center operation. In addition to infrastructure upgrades, operational best practices play a crucial role in reducing Power Usage Effectiveness (PUE). Implementing effective power management policies such as scheduling server maintenance during off-peak hours and utilizing automatic power down features for idle equipment helps minimize unnecessary energy waste. Regular maintenance and preventative checks of cooling systems and power distribution units are vital for maintaining optimal efficiency and preventing energy spikes caused by unexpected equipment failures. Continuous monitoring and analysis of PUE, along with other operational metrics, enable early detection of issues and support proactive adjustments that drive energy optimization.

Importantly, data center efficiency is not just about operational cost savings. Lowering PUE directly reduces the carbon footprint, aligning with the technology industry's increasing focus on sustainability. By adopting strategies that decrease PUE, data center operators help create a greener infrastructure while also cutting energy costs. Improving PUE is an ongoing process, requiring continuous refinement, monitoring, and adaptation to new technologies and changing operational needs. A comprehensive approach—combining physical infrastructure upgrades with improved operational policies is essential for achieving long term energy efficiency and environmental responsibility. More and more, operators are adopting proactive, long term strategies that include sustainable practices throughout the data center lifecycle, supported by ongoing performance analysis and predictive monitoring tools.

Alongside these efforts, load balancing and capacity planning are fundamental to ensuring reliable performance and resource efficiency in modern data centers.

Power Usage Effectiveness (PUE) and Data Center Efficiency Overview

Topic	Description
Total Facility Power vs. IT Equipment Power	Total Facility Power includes all systems (cooling, lighting, PDUs, etc.), as well as IT Equipment. Power includes servers, storage, and networking. PUE = Total / IT Power.
Understanding PUE	A PUE of 1.0 is ideal but unrealistic. A value of <1.2 is excellent, 1.2–1.5 is good, and a value of>1.5 suggests inefficiencies. Climate and facility design affect benchmarks.
Monitoring PUE Over Time	Trends in PUE reveal issues or improvements. Rising PUE may indicate aging systems or inefficiencies; falling PUE reflects optimization success.

Cooling System Efficiency	Outdated cooling increases energy use. Tactics like hot aisle/cold aisle containment, liquid cooling, and free air cooling enhance performance.
Power Distribution Infrastructure	Inefficient or outdated PDUs result in power loss. Upgrading to monitored, load-balanced PDUs improves distribution and cuts waste.
IT Hardware Efficiency	Newer servers with power management features consume less energy. Virtualization and decommissioning idle hardware reduce load.
Operational Practices	Smart capacity planning, proactive maintenance, idle shutdown policies, and anomaly detection drive sustained efficiency.
Real-World Examples	Legacy centers may have PUE >2.0; optimized centers with modern cooling, virtualization, and monitoring can achieve PUE <1.5.
Strategies to Lower PUE	Upgrade cooling and PDUs, improve airflow, adopt virtualization, and phase out obsolete equipment. Aim for a leaner energy footprint.
Power Management Policies	Schedule off-peak maintenance, automate power-down for idle gear, and maintain cooling/power infrastructure to avoid waste.

Sustainability & Environmental Impact	Lowering PUE reduces carbon footprint, aligning with green goals. Efficiency gains support both cost savings and sustainability.
Continuous Optimization	PUE improvement is ongoing—requires infrastructure upgrades, operational discipline, and adaptation to emerging technologies.
Role of Load Balancing and Capacity Planning	Essential for even energy distribution and reliable performance across workloads, aiding overall energy efficiency.

Load Balancing

Effective load balancing distributes incoming traffic and workloads evenly across multiple servers or systems, ensuring no single resource becomes overwhelmed. This enhances application performance, system stability, and utilization efficiency. Different methods exist depending on the architectural needs and scale of the environment:

1. Hardware Load Balancing

This method uses dedicated appliances—commonly known as load balancers or Application Delivery Controllers (ADCs)—positioned between users and servers. These devices apply algorithms like:

- *Round-robin*: Sequentially distributes traffic.
- *Least connections*: Directs traffic to the least loaded server.

- *Source IP hashing*: Routes a client consistently to the same server for session persistence.

  Hardware load balancers are ideal for high traffic environments, offering scalability, health checks, and failover capabilities to maintain uptime and performance.

2. Software Load Balancing

This approach uses programs like HAProxy, Nginx, or Apache to distribute traffic across servers. While more cost-effective and flexible than hardware solutions, their scalability and performance depend on the host server's capacity. These tools continue to support critical features, including traffic shaping, session persistence, and fault tolerance.

3. Application-Level Load Balancing

Tailored for applications with multiple service tiers (e.g., web, application, and database layers), this method directs traffic based on application logic and demand. It is useful for complex workloads requiring session awareness and tier specific balancing strategies.

Successful load balancing is not static it relies on continuous monitoring. Tools like DCIM systems and network performance monitors provide real time visibility into resource use and traffic patterns. Based on this data, administrators can tune configurations and, when needed, leverage **automated scaling** to dynamically provision or decommission servers in response to demand fluctuations.

Capacity Planning

Capacity planning ensures infrastructure can support

current operations and anticipated growth without overprovisioning. A robust planning framework typically follows five key steps:

1. Forecasting Future Demand

Historical data on resource use (CPU, memory, storage, bandwidth) is analyzed using techniques like:

- o Trend analysis
- o Regression modeling
- o Monte Carlo simulations

This forecast must consider business growth, seasonal variations, and future technologies that could impact load.

2. Assessing Current Capacity

A baseline assessment identifies existing infrastructure capabilities and any bottlenecks. Key components evaluated include servers, storage, networking, and environmental systems like cooling and power distribution.

3. Determining Future Capacity Needs

Using forecasts and current assessments, this step calculates the additional infrastructure required, factoring in:

- o Projected growth rates
- o Upcoming application rollouts

o Infrastructure refresh cycles

A safety margin should be included to accommodate unplanned demand surges.

4. **Developing a Capacity Plan**

This plan should define:

o A timeline for acquisition and deployment

o Budget allocation

o Risk assessment and mitigation strategies

It should ensure that the chosen technologies support future scalability and integration with existing systems.

5. Monitoring and Rcvicw

Post implementation, capacity plans must be continuously reviewed. Monitoring tools validate forccasts against rcal-world usagc and identify discrepancies early, allowing for timely adjustments.

Supporting tools include:

• **DCIM (Data Center Infrastructure Management) Systems** provide integrated real time data on resource utilization and environment.

• Performance Monitoring Tools

Track metrics like CPU load, memory, and bandwidth to highlight stress points.

- **Statistical Forecasting Techniques**

  Tools such as exponential smoothing and time series analysis refine demand predictions.

- **Simulation Modeling**

  Enables scenario based planning to test different infrastructure strategies.

Real-World Impact

Organizations that overlook load balancing or fail to plan for capacity risk facing serious consequences. For example, a company experienced outages and performance issues during a major product launch due to underestimating peak demand. Another encountered repeated server failures after centralized traffic overwhelmed its primary application servers. Conversely, businesses that invest in integrated planning and balancing strategies report higher uptime, reduced operational costs, and more scalable systems that adapt smoothly to growth.

By leveraging accurate forecasting, advanced monitoring tools, and adaptive strategies, data center managers can optimize performance and ensure long term infrastructure sustainability. Load balancing and capacity planning are not isolated functions they are key to maintaining service reliability, controlling costs, and supporting business growth. As demand for computing power intensifies, these disciplines will remain central to the evolving mission of critical facilities engineering.

Energy efficiency is crucial in data center operations, greatly affecting both operational costs and environmental sustainability. Reducing energy use without sacrificing performance requires a comprehensive approach, including infrastructure design, equipment choice, and operational practices. This section explores practical strategies to improve energy efficiency in data centers, emphasizing proven techniques and measurable outcomes.

One of the most significant energy consumers in a data center is the cooling system. Optimizing cooling efficiency begins with careful design and selection of appropriate cooling technologies. Traditional methods, such as Computer Room Air Conditioners (CRACs) and Computer Room Air Handlers (CRAHs), while widely used, can often be improved. Modern data centers increasingly leverage more efficient approaches, including:

Free Air Cooling: This technique utilizes ambient air for cooling, eliminating or reducing the need for mechanical cooling systems. This is most effective in climates with consistently cool temperatures and low humidity. Implementing free air cooling involves designing the data center with sufficient ventilation and air filtration systems to ensure proper airflow and prevent the ingress of dust and contaminants. The effectiveness of free air cooling can be significantly enhanced by deploying economizers that automatically switch to free air cooling when environmental conditions are suitable. Proper monitoring of the data center's thermal environment and accurate prediction of ambient conditions are crucial to leverage this strategy efficiently.

Improved Airflow Management: Efficient airflow within the data center is essential for minimizing cooling energy consumption. This involves careful planning of server rack placement, optimizing aisle containment using hot and cold aisle separation, and ensuring adequate clearance for proper air circulation. Implementing raised floor systems can help manage airflow effectively, guiding cooled air to the servers and exhausting hot air separately. Careful consideration of the server rack layout, especially density and orientation, can drastically impact airflow patterns. Tools like Computational Fluid Dynamics (CFD) simulations can assist in visualizing airflow patterns and identifying potential hotspots or areas of poor air circulation. Regular maintenance of airflow systems, including cleaning and replacing filters, is crucial to maintain optimal airflow and prevent performance degradation. Further improvements can be made by using blanking panels to seal empty spaces in racks, further improving airflow efficiency.

Liquid Cooling: Liquid cooling technologies offer a more efficient alternative to traditional air cooling, especially in high-density computing environments. Direct to chip liquid cooling systems directly cool the processors, achieving significantly higher cooling capacity compared to air cooling. Indirect methods like immersion cooling submerge servers in a dielectric fluid, which removes heat more effectively than air. While these methods require a higher initial investment, the long-term energy savings can be substantial, especially with growing server densities. The selection of suitable coolants and the design of the liquid cooling infrastructure require careful consideration of safety, corrosion, and maintenance. Understanding the trade offs between direct and indirect liquid cooling methods is important for selecting the optimal solution for each data center's specific needs.

Beyond cooling system optimization, energy efficient equipment is essential in lowering overall energy use. This involves choosing:

High-Efficiency Power Supplies (PSUs): Modern PSUs with higher efficiency ratings (e.g., 80 PLUS Platinum or Titanium) significantly reduce energy waste. Selecting these PSUs is a relatively straightforward yet impactful step towards energy optimization. The use of power factor correction (PFC) technology in PSUs also contributes to efficiency. Long term energy savings will offset the initial investment in higher-efficiency PSUs. Regular maintenance and periodic testing of PSUs are recommended to ensure their optimal performance and efficiency.

Energy-Efficient Servers and Networking Equipment: Servers and network devices with Energy Star certification meet specific energy efficiency requirements, reducing overall power consumption. The selection of these devices should be integrated into the data center's capacity planning strategy, optimizing for both performance and efficiency. Utilizing virtual machines (VMs) and server virtualization techniques can also enhance energy efficiency by consolidating workloads onto fewer physical servers, reducing idle capacity.

LED Lighting: Replacing traditional lighting with LED options greatly cuts energy use and prolongs the life of the lighting system. The upfront cost of LED lighting is balanced by the significant savings on electricity and lower maintenance expenses. Proper lighting design and placement are crucial to provide adequate illumination and prevent energy waste.

Implementing smart power management strategies further enhances energy efficiency. These include:

Power Usage Effectiveness (PUE) Monitoring and Optimization

Regularly tracking Power Usage Effectiveness (PUE) is essential for understanding and improving overall data center efficiency. As a core metric, PUE provides insight into how effectively power is being used, highlighting inefficiencies in supporting systems like cooling and power distribution. By identifying areas with elevated PUE, operators can pinpoint specific inefficiencies and prioritize improvements. Common interventions include refining airflow management, upgrading cooling technologies, balancing server workloads more efficiently, and optimizing power distribution unit (PDU) performance.

Continuous monitoring—preferably integrated into a DCIM or energy management platform allows operators to track trends over time and respond proactively. For example, a gradual rise in PUE might signal a failing cooling system, misaligned containment strategies, or underperforming infrastructure components. Using real-time analytics and trend analysis, data centers can take timely actions to maintain operational efficiency.

Dynamic Power Management

Dynamic power management techniques offer fine-grained control over energy consumption, adapting usage in real-time to match workload demand. Strategies such as power capping allow operators to set upper limits on server power usage without compromising performance, while power scheduling enables reductions in energy draw during low-utilization periods. When used effectively, these strategies not only reduce energy waste but also extend equipment lifespan by avoiding overuse during off-peak hours.

Successful deployment of dynamic power management relies on accurate demand forecasting and a solid understanding of server workload patterns. Integration with monitoring tools and intelligent automation platforms helps apply real time adjustments based on load projections. This flexible approach contributes to lower PUE and greater cost control while ensuring consistent performance under varying workloads.

Server Consolidation and Virtualization

Virtualization is a foundational strategy in reducing data center energy consumption. By consolidating workloads onto fewer physical servers, organizations can significantly reduce their energy footprint while optimizing space utilization. Virtualization platforms enable efficient workload balancing and allow legacy systems to be retired or repurposed, eliminating power-hungry idle equipment.

Careful design and planning are essential to ensure optimal consolidation ratios without risking performance degradation. This includes analyzing workload behavior, understanding application interdependencies, and monitoring virtual machine (VM) density per host. When properly implemented, virtualization not only lowers energy costs but also simplifies management and improves scalability.

Real-World Impact of Energy Efficiency Initiatives

Numerous data centers have successfully implemented these strategies with tangible results. For example:

- A large scale data center integrated free air cooling with improved airflow containment, reducing annual energy use by 20%.

- Another facility transitioned to next generation energy efficient servers and networking equipment, achieving a 15% reduction in total power consumption.

These cases show that energy efficiency initiatives offer more than environmental benefits they also bring strong financial returns. Many facilities report ROI higher than expected, making these measures both cost effective and sustainable. Success depends on a structured, data driven approach. This involves using energy management platforms with advanced analytics and reporting tools that automate optimization and highlight areas for improvement. Regular energy audits and performance reviews also help ensure ongoing progress and validate further investments in efficiency.

Cost Optimization Beyond Energy Savings

Cost optimization in data center operations extends far beyond energy efficiency alone. A comprehensive strategy includes preventive maintenance planning, vendor management, and strategic technology adoption, all aimed at reducing operational expenses without sacrificing reliability or performance.

Proactive Maintenance Planning

Reactive maintenance—waiting for issues to occur—often results in high emergency repair costs and unscheduled downtime. In contrast, a preventive approach, supported by a Computerized Maintenance Management System (CMMS), ensures regular inspections, cleanings, and component replacements based on manufacturer guidelines and historical performance data.

Advanced analytics and sensor technologies enable predictive maintenance, allowing engineers to detect early signs of equipment degradation. For instance, monitoring the vibration patterns of chillers or the voltage consistency of UPS systems can alert teams to subtle anomalies before they escalate into more significant issues.

A well implemented CMMS supports:

- Maintenance scheduling
- Inventory management
- Cost reporting and optimization
- Prioritization based on failure rates

The result is improved uptime, extended equipment life, and a reduction in unplanned expenditures.

Vendor Contract Optimization

Vendor relationships are another opportunity for cost control. A well-prepared negotiation strategy includes:

- Conducting market research and competitive bidding
- Reviewing historical vendor performance (e.g., SLA compliance, response times)
- Evaluating total cost of ownership (TCO), not just purchase price

Long term contracts with clearly defined service level agreements (SLAs) and warranty terms offer pricing stability and accountability. For example, investing in higher cost servers with better energy efficiency and longer warranties can result in lower TCO than cheaper alternatives with shorter lifespans and frequent maintenance needs.

Building collaborative vendor relationships also improves responsiveness and allows greater flexibility when addressing unforeseen challenges. Periodic reviews and renegotiation ensure that contracts remain favorable as the market and operational needs evolve.

Together, these efforts form a resilient operational framework that enhances performance, lowers expenses, and supports the broader mission of sustainability and uptime assurance in modern data center environments.

Implementing Energy-Efficient Technologies

Adopting energy efficient technologies is a powerful lever for cost reduction in data center operations. Beyond core practices like virtualization and dynamic power management, more advanced strategies can significantly enhance energy performance and reduce operating expenses.

Free cooling techniques, where climate allows, can dramatically reduce reliance on traditional HVAC systems by utilizing outside air or chilled water sources to maintain server room temperatures. When combined with advanced airflow containment such as hot/cold aisle separation or in row cooling—these methods improve temperature regulation and reduce the energy load on cooling systems.

High efficiency Power Distribution Units (PDUs), particularly intelligent or "smart" PDUs, allow granular monitoring and control over energy consumption at the device level. This enables precise allocation of power resources and helps identify underutilized or energy-intensive equipment. When integrated into a centralized Building Management System (BMS), PDUs and other components feed real time data into a unified platform, offering enhanced visibility across the facility.

Data analytics tools further elevate this capability by processing large volumes of energy data to detect patterns of waste and predict future demand. This allows for proactive load balancing, smarter infrastructure utilization, and informed energy policy decisions. LED lighting upgrades, although relatively minor compared to innovations in cooling or power, yield substantial long term savings due to their extended lifespan and low power consumption.

Server consolidation through virtualization, power saving modes for idle equipment, and hardware lifecycle optimization also contribute to minimizing the energy footprint. Many facilities have reported 20% or greater annual energy savings through coordinated energy efficiency programs.

Adopting renewable energy sources like solar or wind power offers long term sustainability advantages while lowering reliance on volatile energy markets. When combined with battery storage and smart load management, renewables can cover a large part of operational power requirements. Many regions also offer tax incentives or rebates, enhancing return on investment. Together, these strategies decrease power consumption and utility costs, as well as reduce the environmental impact of data center operations. Implementing such initiatives also boosts brand reputation—especially among environmentally conscious clients and aligns with corporate sustainability objectives.

Balancing Cost Optimization with Reliability

While pursuing cost savings is important, it should never compromise performance or uptime. A strong infrastructure is the backbone of any operational plan.

Ensuring redundancy in key systems such as power,

cooling, and networking remains vital for business continuity. Backup power systems, dual path networking, and N+1 cooling configurations provide critical fail safes. Skimping on these features for the sake of cost can result in significantly more expensive failures and downtime.

Monitoring and alerting systems play a central role in minimizing risk by enabling early detection of anomalies. A well tuned alerting infrastructure helps prevent minor issues from escalating into full blown incidents.

Investing in high quality equipment often results in a lower total cost of ownership (TCO). While upfront costs may be higher, superior reliability, reduced maintenance frequency, and greater longevity lead to cost savings over time. Choosing the right technology demands a thorough cost benefit analysis, considering not only immediate pricing but also lifecycle costs, performance benchmarks, and support availability.

Regular audits and performance reviews are essential to keep cost saving initiatives aligned with operational goals. These reviews help identify areas where cost measures may have inadvertently impacted reliability and allow for timely course correction.

By tracking KPIs such as:

- Mean Time Between Failures (MTBF)
- Mean Time To Repair (MTTR)
- System Uptime

data center managers gain objective insight into how optimization efforts impact reliability and where adjustments may be needed. A data driven approach enables continuous improvement without compromising stability.

Integrating Sustainability with Cost Efficiency

Cost efficiency and environmental sustainability are not mutually exclusive; in fact, they increasingly reinforce one another. Many modern data centers now treat sustainability as a strategic pillar of their optimization efforts.

Renewable energy adoption is among the most impactful initiatives. Transitioning from fossil fuels to solar, wind, or hydroelectric power significantly reduces greenhouse gas emissions and shields the operation from fluctuations in energy markets. Though initial investment in infrastructure can be high, long term energy savings and available government incentives often make these projects economically viable.

Practical Example:

A data center in a sun rich region installed rooftop photovoltaic (PV) panels matched to its power load profile, achieving a significant reduction in grid reliance. Another site deployed wind turbines after feasibility studies confirmed consistently high wind speeds, further cutting energy bills and emissions.

Site specific feasibility studies are crucial before implementing such technologies. This includes assessing: available land or roof space, solar irradiance or wind maps, load profiles, peak demands, and any possible integration challenges with the current infrastructure.

These projects not only reduce emissions but also **enhance the data center's reputation** and fulfill the sustainability mandates of client organizations.

Water Efficiency and Resource Stewardship

Water conservation is another key sustainability frontier,

especially in facilities that rely heavily on water based cooling.

Modern solutions include:

- **Adiabatic and evaporative cooling:** Significantly reduces water consumption while maintaining cooling efficiency.
- **High-efficiency chillers:** Reduce the water needed per ton of cooling.
- **Leak detection systems:** Prevent water loss and damage.

Regular water usage tracking, aided by automated metering, helps data centers identify wasteful patterns and improve system performance. Some facilities now use closed-loop cooling systems that recirculate water to reduce overall consumption.

Reducing water usage not only lowers operational costs related to utilities and potential fines, but also helps the data center stand out as a responsible corporate citizen, especially in areas facing water scarcity.

In summary, the successful data center of the future will be one that combines cost effective operations, resilient infrastructure, and sustainable practices. Whether through energy monitoring, smart infrastructure upgrades, renewable adoption, or water conservation, these strategies work synergistically to reduce costs and enhance long-term viability. The path to operational excellence lies in data informed decisions, continuous performance tuning, and a commitment to both economic and environmental stewardship.

Chapter 6: Vendor Management and Tenant Interaction

Building and maintaining strong relationships with vendors is paramount to the smooth and efficient operation of a data center. These relationships are the backbone of proactive maintenance, swift troubleshooting, and the overall success of the facility. A well managed vendor ecosystem ensures the timely delivery of services, minimizes downtime, and ultimately protects the critical infrastructure within the data center. This section will explore the key strategies for effectively collaborating with vendors, from the initial selection process to ongoing performance management.

Waste reduction programs are also vital parts of a comprehensive sustainability plan. Data centers produce large amounts of electronic waste (e waste) due to frequent upgrades and replacements of IT equipment. Having a strong e waste management plan, including partnerships with responsible recycling facilities, is essential for proper e waste disposal and minimizing environmental harm. This involves separating different types of e-waste, following safe handling procedures, and ensuring that all materials are properly recycled or reused, thus preventing landfill disposal. Additionally, adopting a responsible purchasing policy that favors equipment with longer lifespans and higher repairability helps decrease the overall e waste produced over time.

Promoting the reuse of older but still functional equipment, when possible, extends the life of assets and decreases the need for frequent hardware replacements. Regular audits of equipment usage can also identify chances to consolidate server resources or adopt virtual machine technologies, thereby lowering the demand for new hardware and helping to reduce waste.

Beyond the immediate environmental impact, sustainability initiatives in data centers significantly reduce operational costs. Lower energy and water consumption can lead to substantial utility savings over time. Additionally, many regions offer financial incentives to encourage the adoption of sustainable practices, further offsetting the capital expenditures associated with these initiatives. These incentives ranging from tax credits and rebates to grants can meaningfully reduce the cost burden of implementing environmentally responsible technologies. The combination of long term cost savings and environmental responsibility makes these strategies particularly attractive to modern data center operators. Moreover, the broader market appeal of environmentally sustainable practices can enhance a company's brand image and attract clients who prioritize corporate responsibility and environmental ethics. To implement effective sustainability programs, data centers must adopt a holistic and strategic approach that goes beyond individual initiatives.

This requires collecting and analyzing detailed data on energy and water consumption patterns, followed by a carefully considered selection of technologies and operational procedures that optimize resource utilization. A multidisciplinary approach integrating environmental science, engineering, and operations—is key to achieving this goal.

Establishing and deploying a comprehensive environmental management system (EMS) is critical for setting clear sustainability goals, measuring performance, and documenting the results of ongoing initiatives. An EMS helps facilitate the tracking of key performance indicators (KPIs) and Critical performance indicators (CPIs) related to energy use, water consumption, waste generation, and greenhouse gas emissions.

This framework enables continuous improvement and ensures accountability, providing a roadmap for ongoing sustainability success. Additionally, it provides a structure for aligning internal practices with external sustainability standards and regulatory requirements, positioning the data center as a leader in environmental responsibility.

Quantifying the success of sustainability initiatives depends on the systematic tracking of relevant metrics. CPIs such as Power Usage Effectiveness (PUE), Water Usage Effectiveness (WUE), and Carbon Footprint (measured in tons of CO2 equivalent) serve as essential tools for evaluating performance. PUE measures overall energy efficiency by comparing total facility power consumption with the power used by IT equipment alone, while WUE provides insight into water resource efficiency. Carbon Footprint calculations, meanwhile, help identify the total emissions generated by a facility's operations. Regular review and reporting of these CPIs and KPIs make it possible to identify inefficiencies, assess the impact of implemented strategies, and continuously refine approaches to achieve better results. Transparency in reporting not only builds trust with stakeholders but also signals a strong commitment to environmental stewardship. Sharing data with industry peers further promotes best practices and can accelerate the adoption of sustainable technologies and policies across the data center sector as a whole.

Incorporating sustainability into core operations is not merely an ethical choice but a strategic business decision that delivers both environmental and financial benefits. By integrating renewable energy sources, applying advanced water conservation techniques, and instituting comprehensive waste reduction measures, data centers can significantly minimize their environmental footprint and improve long-term cost efficiency. A comprehensive strategy—one that includes the right technologies, accurate data analysis, and an ongoing commitment to monitoring and improvement ensures that sustainability initiatives are not only implemented but sustained. Data centers must remain agile and proactive, adapting to evolving technologies and emerging sustainability opportunities. Those that do will be better positioned to meet growing environmental expectations, comply with regulatory requirements, and capture market share in a world increasingly driven by environmental considerations.

In this context, sustainability is not just a set of actions it is a forward looking framework that supports operational excellence, long term cost control, and responsible innovation in the digital infrastructure space.

Vendor performance management involves continuous monitoring and assessment of the vendor's services. Critical performance indicators (CPIs) are crucial tools for tracking vendor performance against predefined targets. These metrics can include response time, resolution time, availability, and adherence to SLAs.

Regular reports summarizing vendor performance should be generated and reviewed to ensure that the vendor consistently meets the required standards. Consistent monitoring enables proactive intervention to address any performance issues before they escalate into major problems. Regularly scheduled meetings with vendor representatives allow for proactive problem solving and the identification of potential issues. A system for tracking and documenting all interactions with the vendor, including any issues encountered and their resolution, is invaluable for future reference.

Vendor Management and Sustainability Overview

Topic Description	
Vendor Relationships	Strong vendor ties support proactive maintenance, timely service delivery, and minimal downtime. Selection and performance management are key.
Vendor Performance Management	Use KPIs/CPIs (e.g., response time, SLA adherence, resolution time) to monitor performance. Maintain logs, conduct regular reviews, and meetings.
Electronic Waste (E-Waste) Management	Implement responsible recycling programs, separate and safely handle e-waste, and partner with certified recyclers.
Equipment Reuse & Purchasing	Prioritize repairable, long-life hardware. Reuse older yet functional devices and consolidate IT resources where possible.
Sustainability and Cost Savings	Reducing energy and water use lowers operational costs. Financial incentives (tax

	credits, rebates) can offset sustainability investments.
Corporate Responsibility Benefits	Sustainability boosts brand image, appeals to environmentally conscious clients, and supports regulatory compliance.
Strategic Sustainability Approach	Use detailed resource data to drive decisions. Align environmental science, engineering, and operations for maximum impact.
Environmental Management System (EMS)	A formal EMS tracks CPIs, aligns practices with regulations, supports continuous improvement, and sets sustainability goals.
Key Sustainability KPIs/CPIs	Track PUE (energy efficiency), WUE (water efficiency), and Carbon Footprint (CO_2 emissions) to measure and optimize performance.
Data Transparency and Reporting	Regularly report sustainability metrics to build trust, support industry benchmarking, and encourage broader green adoption.
Integrated Operations and Technology	Combine renewable energy, water conservation, and waste reduction into core operations. Stay agile to adopt emerging technologies.

Sustainability as Strategy	Sustainability supports innovation, operational excellence, regulatory compliance, and long-term market advantage.

Addressing performance issues requires a structured and collaborative approach. When a vendor does not meet its obligations, a clear and documented escalation process should be followed. This usually starts with informal discussions and escalates to more formal written communications if the issue remains unresolved. The vendor should be given clear expectations and opportunities to fix the deficiencies. Performance Improvement Plans (PIPs) can be created to specify the actions the vendor needs to take to resolve the problems. Regular follow-up meetings help monitor progress and ensure the vendor complies with the PIP. Continued failure to meet agreed standards may lead to contract termination or switching to alternative vendors. The entire process should be thoroughly and properly documented.

Building strong relationships with vendors extends beyond transactional interactions. Proactive relationship building cultivates mutual trust and respect.

This involves actively engaging with vendor representatives, attending industry events, and sharing best practices to foster collaboration and innovation. Collaborative problem-solving fosters a shared sense of purpose, leading to more innovative and efficient solutions. A strong vendor relationship goes beyond mere contractual obligations; it involves partnership and commitment to shared success. Such collaborations can involve joint training sessions to enhance the data center team's understanding of the equipment or systems managed by the vendor, improving overall responsiveness to issues. Regular communication, even outside of immediate problem situations, builds rapport and reinforces the collaborative nature of the relationship. Treating vendors as partners, not just contractors, fosters loyalty and encourages them to invest in the success of the data center.

In conclusion, effective vendor management is critical to the success of any data center operation. A well defined selection process, carefully negotiated contracts, consistent performance monitoring, and open communication are essential elements of successful vendor relationships. By adopting these practices, data center engineers can build strong, collaborative partnerships that ensure the reliable and efficient operation of their critical infrastructure.

The resulting positive relationship with your vendors will contribute significantly to reduced downtime, enhanced efficiency, and ultimately, a more robust and resilient data center environment. Proactive vendor management is an investment in the long term stability and success of the data center. It's a crucial skill for critical facilities engineers and a key component of ensuring the ongoing operation of the data center. This strategic approach reduces risks, minimizes disruptions, and contributes to a more efficient and reliable data center operation. The proactive management of vendor relationships is an investment that yields significant returns in terms of improved performance, reduced costs, and a more robust and dependable data center operation. Understanding the intricacies of Service Level Agreements (SLAs) is crucial for successful data center operations. SLAs serve as the cornerstone of the relationship between the data center operator and its tenants, clearly defining expectations, responsibilities, and consequences for both parties.

A well defined SLA provides a framework for consistent service delivery, mitigating potential disputes and ensuring a harmonious working relationship. This understanding extends beyond a simple document; it represents a commitment to mutual success and shared accountability.

The creation of a robust SLA begins with a thorough understanding of the tenant's specific requirements. This involves detailed discussions to ascertain their critical business needs and the level of service they require. Factors such as required uptime, response times for outages, and recovery time objectives (RTOs) are paramount in this process. For example, a financial institution with high frequency trading operations will have drastically different SLA requirements compared to a smaller business using the data center for basic web hosting. The criticality of the tenant's operations directly influences the specifics of the SLA.

Once the tenant's needs are clearly defined, the data center operator can begin crafting the SLA document. This document should be clear, concise, and unambiguous, avoiding technical jargon that might be misinterpreted. It needs to explicitly state the agreed upon performance metrics, including uptime guarantees (often expressed as a percentage of annual uptime), response times for incidents, and resolution times for outages. For instance, an SLA might specify a 99.99% uptime guarantee, with a maximum response time of 15 minutes for critical incidents and a maximum resolution time of four hours for those same incidents. These metrics must be measurable and verifiable, allowing for objective assessment of performance.

Key Performance Indicators (KPIs) and Critical Performance Indicators (CPIs) are the lifeblood of any effective

SLA. These metrics serve as a yardstick against which performance is measured. Common KPIs/CPIs include mean time to repair (MTTR), mean time between failures (MTBF), and availability. The SLA should clearly define which KPIs/CPIs are being tracked and the target values for each. It's vital to establish a robust monitoring system that captures this data accurately and reliably.

This system should provide real time insights into the data center's performance, allowing for proactive identification and mitigation of potential issues before they impact the tenant's services. Regular reporting on these KPIs/CPIs is crucial for transparency and accountability. This reporting should be clearly structured and easily understood, presenting data in a concise and readily interpretable format.

Reporting procedures form a critical part of the SLA. These procedures must outline how performance data will be collected, analyzed, and presented to the tenant. Regular reports, typically monthly or quarterly, should be generated to track performance against the agreed upon KPIs/CPIs. These reports should be readily accessible to the tenant, providing transparency and building trust. Moreover, the reporting process should outline the methods for handling exceptions or incidents that might affect performance. This might involve detailed incident reports that document the cause of an outage, the actions taken to resolve it, and the impact on the tenant's operations. A well defined reporting process ensures that both the data center operator and the tenant are kept informed of the overall health of the infrastructure and the services being provided.

A critical aspect of managing SLAs is the establishment of a clear process for addressing SLA breaches. When performance falls below the agreed upon levels, the SLA should outline a specific procedure for handling the situation.

This might involve escalation procedures, notifications to designated personnel, and agreed upon remedies for the disruption. These remedies might include financial credits, service restoration guarantees, or other forms of compensation, all pre defined within the SLA.

The process should be transparent and fair, aiming to resolve the issue promptly and effectively while maintaining a positive tenant relationship. Documentation is essential throughout this process, providing a clear record of the breach, the steps taken to resolve it, and the agreed upon remedies.

Maintaining a positive tenant relationship is paramount, even in the event of an SLA breach. Open and honest communication is crucial in these situations. Proactive communication from the data center operator, keeping the tenant informed of the situation and the steps being taken to resolve it, can significantly mitigate any negative impact on the relationship. Transparency, by openly sharing information and acknowledging any shortcomings, builds trust and fosters a collaborative spirit. Regular review meetings, both scheduled and ad hoc, facilitate open dialogue and problem solving, ensuring the SLA remains a living document that is adaptable to evolving needs and technological advancements. Consider a scenario where a tenant experiences an unexpected surge in traffic, resulting in brief periods of performance degradation. A well defined SLA will outline a clear process for handling this situation, perhaps allowing for temporary exceptions or adjustments to the performance metrics during such unforeseen events. Conversely, a poorly defined SLA might lead to disputes and strained relationships, hindering the overall efficiency of the data center.

Another example would involve a prolonged outage caused by a critical equipment failure. The SLA should clearly outline the data center's response time, recovery time objectives, and potential compensation for any disruption. A clear and well documented process ensures that both parties understand their responsibilities and obligations, minimizing potential conflicts and maintaining a constructive working relationship.

In conclusion, SLAs are much more than just contractual obligations; they form the foundation of a successful partnership between data center operators and their tenants. By carefully designing and managing SLAs, including robust CPI reporting and clear breach resolution procedures, data center operators can build trust, ensure consistent service delivery, and foster long-term, mutually beneficial relationships. Proactively managing SLAs is essential for efficient data center operation and contributes directly to the facility's overall success and stability. Regularly reviewing and adapting SLAs based on operational experience and changing tenant needs will help maintain their relevance and effectiveness in protecting relationships and ensuring quality service. Paying close attention to these factors creates a flexible and responsive framework that encourages collaboration, promotes transparency, and guarantees the high-quality service delivery needed for modern business operations. This goes beyond just avoiding litigation; it's about nurturing a partnership rooted in mutual trust and a shared goal of success.

Effective communication and collaboration with tenants are paramount to successful data center operation. This goes beyond simply responding to requests; it's about proactively building relationships, fostering trust, and ensuring a mutually beneficial partnership.

Proactive communication prevents misunderstandings and allows for early identification and resolution of potential issues before they escalate into major problems.

One effective strategy is to establish regular communication channels. This might involve scheduled meetings, perhaps monthly or quarterly, to discuss performance, upcoming maintenance, and any planned changes to the data center infrastructure. These meetings provide a platform for open dialogue, allowing tenants to voice concerns, ask questions, and provide feedback. Furthermore, establishing a dedicated point of contact for each tenant ensures efficient and streamlined communication. This person acts as a single point of accountability, responsible for addressing tenant inquiries and coordinating responses from various departments within the data center. Such designated contact persons should be empowered to make decisions and solve problems independently, eliminating delays caused by multiple layers of approval.

Proactive communication also involves informing tenants of planned maintenance activities well in advance. This allows tenants to plan accordingly, minimizing potential disruptions to their operations. Clear communication concerning the scope of the maintenance, its duration, and any potential impact on their services is essential. Providing detailed notifications, which may include the use of email alerts, SMS notifications, or even dedicated tenant portals, ensures that everyone is well informed and prepared. It's also beneficial to offer alternative solutions where possible, for example, providing access to redundant systems or resources to ensure business continuity during planned outages.

Beyond scheduled communications, it's critical to establish a responsive system for addressing urgent issues and inquiries. This often involves a multi-tiered support system, handling requests and issues based on their urgency and complexity. A 24/7 help desk, with readily available personnel to respond to urgent issues, ensures uninterrupted support. This help desk should have the capability to escalate urgent issues to the appropriate engineering teams rapidly. A ticketing system that tracks and manages requests and incidents, providing updates and status reports, ensures transparency and promotes accountability. This system provides a valuable audit trail, documenting the history of all interactions and facilitating efficient problem solving.

Effective communication involves using multiple communication channels, tailored to the specific needs and preferences of each tenant. This may involve email for routine updates, phone calls for urgent matters, and video conferencing for complex discussions. The selection of the appropriate channel depends on the urgency of the communication and its subject matter. The goal is to provide timely and relevant information, using a channel that is both efficient and effective. Additionally, communication should be adapted to the level of technical understanding of the tenant. While technical jargon is sometimes necessary, it's important to ensure that all communication is clear, concise, and readily understandable by all parties. Avoiding technical jargon whenever possible and providing plain language summaries helps ensure that messages are universally understood and prevent misinterpretations.

When addressing tenant issues, a structured approach is critical. This starts by actively listening to the tenant's concerns, acknowledging their perspective, and expressing empathy. It's crucial to understand the full impact of the issue on their operations. A detailed investigation into the problem, including the collection of relevant data and information, must follow. This investigation might involve the examination of system logs, network performance data, or even on site inspections. Based on the findings of this investigation, a comprehensive resolution plan should be developed and communicated to the tenant. This plan should outline the steps that will be taken to resolve the issue, a timeline for the completion of these steps, and the anticipated impact on the tenant's operations.

Maintaining transparency throughout the process is crucial. Regular updates should be given to the tenant, informing them of progress and any changes to the resolution plan. Open communication about potential delays or unexpected challenges helps sustain a collaborative environment and avoids misunderstandings. Once the issue is resolved, a follow up message should be sent to confirm the resolution and ensure the tenant's satisfaction. Gathering feedback from the tenant about the resolution process is essential for continuous improvement and better future responses.

Building rapport with tenants is essential for effective tenant interaction. This includes demonstrating professionalism, competence, and a genuine interest in the tenant's success. Consistent interaction, beyond just fixing issues, helps build positive relationships. Regular informal check ins, casual chats, and an open-door policy all contribute to a more cooperative atmosphere. Hosting informal events, like networking opportunities or social gatherings, can further strengthen these bonds and create a more positive and collaborative environment.

Addressing tenant concerns promptly and effectively is critical. Delays can lead to frustration, and prompt resolutions demonstrate professionalism and commitment to service. A well defined escalation path helps ensure that issues are addressed promptly, with escalation to higher levels of management when necessary. Having a clear protocol for managing escalated incidents, including clearly defined roles and responsibilities, streamlines the process and ensures efficient resolution. This also helps foster a positive working relationship with the tenant, as they know their concerns are taken seriously and dealt with efficiently.

Consider a scenario where a tenant experiences an unexpected network outage. Effective communication involves immediately notifying the tenant, providing an initial assessment of the issue, and outlining the steps being taken to resolve it. Regular updates, providing an estimated time of restoration, should be provided.

Following the restoration, a thorough post incident review should be conducted to determine the root cause and to implement preventive measures. Transparency throughout the process, acknowledging any shortcomings and explaining the steps taken to prevent similar occurrences in the future, fosters trust and strengthens the tenant relationship.

Another example involves a tenant requesting a change to their existing infrastructure. Effective communication involves a collaborative discussion to determine the feasibility of the request, to explore alternatives, and to develop a detailed implementation plan. This plan would outline the steps involved, the required timelines, and any potential impact on other tenants or the overall data center infrastructure. Open communication and transparency throughout the process ensure mutual understanding and minimize potential disruptions. Careful planning, with a clear understanding of the tenant's needs and the data center's capabilities, is essential.

In conclusion, effective tenant communication and collaboration are pivotal to the success of any data center operation. It's not merely about reacting to issues but about proactively building relationships, fostering trust, and ensuring that the data center consistently meets and exceeds tenant expectations. By implementing the strategies outlined above, data center operators can cultivate a collaborative environment, resolve issues efficiently, and ultimately strengthen their relationships with their tenants, resulting in a mutually beneficial and enduring partnership. This proactive approach contributes to the stability and success of the data center as a whole, demonstrating that a well managed tenant relationship is as crucial to operation as any technical system.

SLA Management and Tenant Communication Overview

Topic Description	
Service Level Agreements (SLAs)	SLAs define response times, recovery objectives, and compensation terms. Well-documented SLAs clarify responsibilities and minimize conflict.

SLA as a Strategic Tool	More than contracts—they build trust, ensure reliable service delivery, and support long-term tenant relationships through proactive management.
SLA Review and Adaptation	Regularly update SLAs based on operational changes and tenant needs to keep agreements relevant and effective.
Proactive Tenant Communication	Regular updates and open dialogue help prevent issues. Foster trust through proactive, transparent communication strategies.

Scheduled Tenant Meetings	Hold monthly or quarterly meetings to discuss maintenance, performance, and infrastructure changes. Encourage feedback and collaborative planning.
Dedicated Tenant Contacts	Assign a point of contact per tenant to streamline issue resolution and communication. Ensure they're empowered to act quickly.
Planned Maintenance Notifications	Notify tenants in advance via email, SMS, or portals. Include scope, duration, impact, and contingency plans to reduce service disruption.

Urgent Support Systems	Implement 24/7 help desks, tiered support, and escalation paths. Track issues with ticketing systems for transparency and accountability.
Multichannel Communication	Use email for updates, calls for urgent issues, and video conferencing for complex topics. Tailor messages to each tenant's needs.
Clear, Understandable Messaging	Avoid jargon where possible. Use plain language to ensure technical and non-technical stakeholders understand the message.
Structured Issue Resolution	Listen actively, investigate with data, and provide a resolution plan with clear steps and timelines. Communicate progress regularly.
Post-Issue Follow-Up	Confirm resolution, gather feedback, and use insights to improve future responses. Transparency builds long-term trust.
Relationship Building	Foster rapport through informal check-ins, open-door policies, and occasional social events to build a collaborative atmosphere.
Efficient Escalation Procedures	Define clear escalation paths, assign roles, and ensure the timely resolution of high-impact issues to maintain trust and service levels.

Incident Communication Example	During a network outage, immediately notify tenants, provide updates, outline steps taken, and conduct a transparent post-incident review.
Infrastructure Change Requests	Collaborate on feasibility and planning, communicate timelines and impacts, and ensure smooth implementation through clear communication.

Change management within a data center environment is not merely a set of procedures; it's a critical operational philosophy that directly impacts uptime, efficiency, and the overall success of the facility. Effective change management minimizes disruptions, prevents costly system failures, and fosters trust with both tenants and vendors. The absence of robust change management processes can lead to cascading failures, prolonged downtime, and significant financial losses. This section will delve into the importance of meticulously planned and meticulously executed change management, encompassing all stages from initial request to post-implementation review.

The foundational element of any successful change management process is a clearly defined and well documented procedure. This procedure should outline the steps involved in initiating, reviewing, approving, implementing, and closing out a change. The procedure must be accessible to all relevant personnel, including engineers, technicians, vendors, and tenants. Furthermore, it should be regularly reviewed and updated to reflect changes in technology, operational procedures, and industry best practices. This ensures the procedure remains relevant and effective in addressing the evolving needs of the data center. The documentation should include detailed forms, templates, and checklists to streamline the process and minimize the risk of errors.

The change request process is the initial and vital step. A standardized change request form should be used for all modifications, regardless of their size or complexity. This form should include a comprehensive description of the proposed change, its purpose, the potential impact on other systems, required resources, and a detailed implementation plan. The level of detail needed will vary based on the complexity and risk involved with the change. For example, a simple software update might need less thorough documentation compared to a major hardware upgrade or a large network reconfiguration. In every case, a clear justification for the change must be included, explaining the business necessity or technical requirement behind it.

Once a change request is submitted, it undergoes a thorough review process. This process typically involves multiple levels of approval, depending on the scope and potential impact of the change. Minor changes might only need approval from a single engineer or team lead. However, significant changes that require extensive resources or affect critical systems will need approval from senior management and possibly external stakeholders such as tenant representatives. This multi tiered approval process ensures that the risks linked to the change are carefully evaluated and controlled before proceeding. Each approval level should be documented, along with any conditions or requirements imposed as part of the approval.

The implementation phase is crucial. Before making any changes, a detailed risk assessment should be conducted to identify potential issues and develop strategies to address them. This assessment should evaluate the potential effects on all systems and services within the data center, including tenant operations. A comprehensive test plan should be created and carried out before deployment to ensure the change works as intended and does not adversely affect other systems. This plan should include both unit testing (testing individual components) and integration testing (examining how components work together). Testing should be thoroughly documented, including results and any problems found.

Effective communication is essential during the implementation phase. Tenants should be informed in advance of any changes that could impact their operations, including estimated downtime or service interruptions. This advance notice helps tenants prepare contingency plans and reduces disruptions to their business activities. Regular updates should be provided throughout the implementation process to keep tenants informed of progress and address any concerns. Clear and concise communication is vital to maintaining trust and confidence. This may include email updates, phone calls, or dedicated tenant portals, depending on the complexity and impact of the change.

Post implementation review is a often overlooked but essential step in the change management process. After the change has been implemented and verified, a comprehensive review should be conducted to evaluate its effectiveness and learn from the experience. This review should include an analysis of the implementation process, such as the time taken, resources used, and any unexpected issues encountered. It should also assess the impact of the change on the overall performance of the data center and tenant satisfaction. Feedback should be collected from all involved parties, including engineers, technicians, vendors, and tenants. This feedback is vital for continuously improving the change management process and avoiding similar issues in the future. This post-implementation review serves as the foundation for ongoing refinement of the change management process, making it more efficient, reliable, and less error prone.

Coordination with vendors is a significant aspect of successful change management. Vendors often play a key role in implementing changes, whether it's installing new hardware, upgrading software, or performing maintenance. Effective coordination involves establishing clear communication channels and responsibilities. This includes providing vendors with sufficient lead time to plan and execute their tasks, ensuring they have access to the necessary information and resources, and establishing clear escalation paths for addressing any issues that may arise. Regular meetings and progress reports help ensure that the vendor's work is aligned with the overall change management plan. A documented Service Level Agreement (SLA) with the vendor should clearly define their responsibilities, timelines, and performance expectations, providing a framework for accountability and dispute resolution.

Consider a scenario where a major network upgrade is planned. The change management process would begin with a detailed request, outlining the specific upgrade, its justification, and its potential impact on tenants. The request would be reviewed and approved by the appropriate personnel. Then, a detailed implementation plan would be developed, outlining the specific steps involved, the resources required, and a detailed timeline. This plan would include communication strategies for informing tenants of the planned outage and providing regular updates. After implementation, a thorough testing phase would validate the functionality of the new network. Finally, a post implementation review would analyze the entire process, identifying areas for improvement in future upgrades.

Another example could be a tenant requested change. Perhaps a tenant needs to increase their power capacity. The change request would detail the additional power requirements, and the feasibility would be assessed. This might involve load balancing studies to ensure the data center can accommodate the additional load without compromising redundancy or stability. The change would be carefully planned, possibly requiring work outside of normal operational hours to minimize impact. The tenant would be kept fully informed throughout the process, from initial request approval to final completion, emphasizing transparency and collaboration. Following the successful completion, a post implementation review would analyze the efficiency of the process and its impact on overall data center operations.

In conclusion, a robust change management process is not merely a set of forms and procedures; it's a cornerstone of reliable data center operation. By meticulously planning, implementing, and reviewing changes, data center engineers can minimize disruptions, prevent failures, and foster positive relationships with tenants and vendors. This proactive approach, coupled with transparent communication, ensures the continued success and stability of the data center, demonstrating that a well managed change process is as essential as the technical systems themselves.

Managing tenant expectations is crucial for maintaining a successful and harmonious data center environment. It's not just about fulfilling contractual obligations; it's about building a collaborative relationship based on trust and transparency. Proactive communication, realistic service level agreements (SLAs), and a responsive approach to tenant concerns are essential parts of effective tenant expectation management. One of the best strategies is setting clear expectations from the beginning. Before a tenant moves into the data center, a thorough onboarding process should be put in place. This process should clearly outline the services offered, the SLAs in effect, and the procedures for reporting and resolving incidents. This upfront clarity reduces misunderstandings and potential conflicts later on. The onboarding should include a detailed tour of the facility, highlighting key infrastructure components and emergency procedures.

Tenants should be introduced to key personnel within the data center team, establishing clear communication channels. A well documented tenant handbook, easily accessible both online and offline, is invaluable in reinforcing this initial information and providing a consistent reference point. This handbook should include details on power usage, cooling capacity, network connectivity, security protocols, and maintenance schedules. Regular updates to this handbook reflect changes in the data center's capabilities and operational procedures.

A key part of setting expectations is establishing realistic SLAs. Overpromising can cause dissatisfaction and strain relationships. SLAs should be based on a thorough evaluation of the data center's capabilities and include specific metrics for uptime, response times, and resolution times. These metrics need to be clearly defined and easily understood by both the data center staff and tenants. Importantly, SLAs should be realistic and reflect the data center's actual capacity and operational strengths. Inflated SLAs might seem attractive to win contracts, but they can lead to conflicts later. It's better to deliver reliable service within true capabilities than to make promises that cannot be kept. A realistic SLA builds trust, and tenants will value transparency and accuracy over unmet expectations.

Regular communication with tenants is crucial in maintaining positive relationships. This doesn't simply mean responding to incidents or requests; it's about proactively engaging with tenants and keeping them informed about the data center's performance and any planned maintenance or upgrades. Regular newsletters, updates on the facility's management system, and open communication channels allow the facility team to proactively share relevant information. This proactive communication keeps tenants informed and reduces the possibility of surprises, which can lead to heightened tension and distrust. This regular interaction might involve scheduled meetings, perhaps quarterly or biannually, to discuss any concerns, planned upgrades, or upcoming changes that could affect tenants' operations.

For example, a planned electrical system upgrade might require a scheduled downtime. Proactive communication, provided several weeks in advance, outlining the scope of work, estimated downtime, and contingency plans, allows tenants to prepare accordingly, minimizing disruption to their business operations. This proactive approach to communication is far more effective than a reactive approach, which only addresses the problem after it has occurred, potentially leaving tenants feeling frustrated and unsupported. Effective communication involves using multiple channels email, phone calls, and potentially tenant portals to ensure that information is consistently relayed and received.

Transparency is another key aspect of managing tenant expectations. Openly sharing information about the data center's performance, including metrics such as power usage effectiveness (PUE), uptime, and incident response times, builds trust and fosters a sense of collaboration. Regular reporting, including performance reports and incident reports, should be provided to tenants, allowing them to monitor the data center's performance and identify any potential issues. This transparency demonstrates that the data center team is committed to accountability and is working towards a mutually beneficial partnership. This open sharing of information enables a more collaborative working relationship. Access to real time monitoring dashboards, if technologically feasible and secure, allows tenants to visualize the data center's performance and feel more directly connected to its operations.

When addressing tenant concerns, a quick and professional reply is essential. Having clear procedures for escalation ensures that issues are handled quickly and effectively, no matter how complex they are. When a tenant reports a problem, the data center team should acknowledge it right away, explain the steps being taken to fix it, and give regular updates on the progress. This method shows the team's commitment to good support and helps build trust with tenants. It's important to avoid giving vague or misleading information; honesty is key to keeping a strong relationship. Providing accurate and timely info helps prevent misunderstandings, which often cause conflicts.

Additionally, conflict resolution procedures must be clearly outlined. This includes steps for managing disputes related to SLAs, service performance issues, billing disagreements, or other conflicts. An explicit escalation process is essential, ensuring issues are directed through various management levels when needed. A mutually agreed upon method for resolving disputes, like mediation or arbitration, can help settle conflicts fairly and efficiently. This prevents minor problems from turning into major conflicts that could harm the relationship. Regularly reviewing the conflict resolution procedures helps ensure they stay relevant and effective.

Consider a scenario where a tenant experiences a prolonged power outage. An effective response would involve immediately acknowledging the outage, identifying the root cause, implementing corrective actions, and providing frequent updates to the tenant on the progress of restoring power. Transparent communication throughout the process, even if it involves sharing information about the investigation's ongoing nature, builds confidence. A post incident review, shared with the tenant, details the root cause analysis and the corrective actions taken to prevent future outages. This demonstrates accountability and a commitment to continuous improvement. The review may involve the tenant and could highlight areas for both the tenant and the data center to improve their respective processes. This approach strengthens the relationship rather than creating mistrust or animosity.

Another example: a tenant reports slow network connectivity. A prompt response involves immediate investigation, identifying the source of the problem whether it's an issue with the tenant's equipment, the data center's network infrastructure, or external network issues.

In summary, managing tenant expectations involves a multifaceted approach. It requires proactive communication, realistic SLAs, transparency, and a prompt and professional response to tenant concerns. By consistently demonstrating reliability, professionalism, and a commitment to collaboration, data center engineers can build strong relationships with tenants, fostering a successful and mutually beneficial partnership. This focus on strong tenant relationships creates a more stable and productive data center environment for everyone. The effort invested in proactively managing tenant expectations pays off significantly in reduced conflicts, increased tenant satisfaction, and a more harmonious data center ecosystem.

Acknowledgement

I would like to express my sincere gratitude to several individuals and organizations who contributed significantly to the completion of this book. First and foremost, I want to thank Gabriel Jushua, who has been a great teacher. I also thank Danny Lopez, District Training Engineer, whose expertise provided invaluable insights and real world examples throughout the process. Their dedication to operational excellence within the critical facilities environment was instrumental in shaping the practical, hands on approach that defines this guide.

Finally, I am deeply grateful to my family for their unwavering patience and support during the long hours dedicated to writing and researching this project. Their understanding and encouragement were invaluable.

Appendix

Appendix A: Sample Service Level Agreement (SLA) Template

Appendix B: BMS System Alarm Codes and Troubleshooting Guide (includes a table detailing common BMS alarm codes, their likely causes, and suggested troubleshooting steps)

Appendix C: Lockout/Tagout Procedure Checklist (a detailed checklist for safe and effective lockout/tagout procedures)

Appendix D: Emergency Response Plan Template (a template for creating and implementing a comprehensive data center emergency response plan)

BMS: Building Management System – A centralized system for monitoring and controlling various building systems, including HVAC, power, security, and fire protection.

CDU: Computer Room Air Handling Unit – A more versatile unit that can handle both cooling and heating.

CPI: Critical Performance Indicator – A vital metric used to measure the most essential activities or outcomes that determine the success or failure of a project, business, or organization.

CRAC: Computer Room Air Conditioner – A specialized air conditioning unit designed for data centers.

HVAC: Heating, Ventilation, and Air Conditioning – The systems responsible for maintaining the appropriate temperature and humidity levels within a data center.

KPI: Key Performance Indicator – A measurable value that demonstrates how effectively an individual, team, or organization is achieving specific objectives.

PUE: Power Usage Effectiveness – A metric used to measure the efficiency of a data center's power usage. A lower PUE indicates greater efficiency.

SLA: Service Level Agreement – A contract outlining the agreed upon service levels between a data center provider and its tenants.

UPS: Uninterruptible Power Supply – A device that provides backup power to critical equipment during power outages.

Other relevant terms

ASHRAE Standard 90.1: Energy Standard for Buildings Except Low Rise Residential Buildings.

BICSI Data Center Design and Implementation Best Practices.

Uptime Institute Data Center Tier Standard: Topology.

Made in United States
North Haven, CT
10 November 2026

22091330R00098